WRITING OPERATING PROCEDURES FOR PROCESS PLANTS

Ian S. Sutton

Second Edition, Completely Revised

Southwestern Books, Houston, Texas, United States of America

WRITING OPERATING PROCEDURES FOR PROCESS PLANTS

ISBN 1-57502-057-2
Library of Congress Cataloging-in-Publication Data
95-71131

Sutton, Ian S.
Writing Operating Procedures For Process Plants/
Ian S. Sutton

Published and distributed by Southwestern Books, Houston, Texas

Southwestern Books, PO Box 58960-195, Houston, Texas, United States of America, 77258-8960.

Fax: (713) 488-2259

Printed in the USA by

MORRIS PUBLISHING
3212 E. Hwy 30
Kearney, NE 68847
800-650-7888

WARNING — DISCLAIMER

This book is designed to provide information in regard to the subject matter covered. It is sold with the understanding that the publisher and author are not engaged in rendering legal, accounting or other professional services. If legal or expert assistance is required, the services of a competent professional should be sought.

It is not the purpose of this book to reprint all the information that is available to the author and/or publisher, but to complement, amplify and supplement other texts. You are urged to read all the available material, learn as much as possible about Operating Procedures and Process Safety Management, and to tailor the information to your needs. For more information, see the references provided at the end of this book.

Every effort has been made to make this book as complete and as accurate as possible. However, there **may be mistakes**, both typographical and in content. Therefore, this text should be used only as a guide and not as the ultimate source of information and guidance on Operating Procedures and Process Safety Management. Furthermore, information on these topics will have changed since the printing date of this book.

The examples in this book are fictional, and have been created specifically for this book. The book does not contain information about any specific plant, process, company or individual. Nor does it describe the policy of any company.

The purpose of this book is to provide information in the above areas. The author and publisher of this book shall have neither liability nor responsibility to any person or entity with respect to any loss or damage caused, or alleged to be caused, directly or indirectly by the information and/or guidance provided in this book.

If you do not wish to be bound by the above, you may return this book to the publisher for a full refund.

Additional Materials

Training Courses on the materials in this book are available. To learn more, please contact Southwestern Books.

A booklet entitled *Operating Procedures For Process Plants — Writer's Guide and Worked Examples* is also available. This booklet provides a variety of examples that can be used to give you ideas about how to write your procedures and Writer's Guide.

If you are responsible for developing an Operating Procedures program for your company or your client, send for *Operating Procedures For Process Plants — A Leader's Guide.* This is a set of course materials, overhead transparencies (many in color) that you can use to lead your colleagues in discussions on this topic. The *Leader's Guide* is based on the materials provided in this book.

About The Author

Ian S. Sutton is a chemical engineer with over 20 years of diversified experience in the process industries. He is a principal engineer with Fluor Daniel Inc. in Houston, Texas. Mr. Sutton specializes in Process Safety Management, with a particular emphasis on the development of operating manuals. He is a professional engineer registered in the State of Texas.

TABLE OF CONTENTS

Chapter 2 — Volume I - The Operating Procedures

Chapter 3 — Volume II - Operating Information

Chapter 4 — Volumes III/IV - Emergency & Troubleshooting Procedures

Chapter 5 — Volume V - The Initial Start-Up Manual

Chapter 6 — Writing and Publishing The Procedures

Chapter 7 — Managing A Procedures Writing Project

Chapter 8 — Process Safety Management and Operating Procedures

Figures

Tables

PREFACE

More and more companies in the process industries are facing up to the task of writing operating procedures. For many of them, the immediate driving force comes from a government regulation, often OSHA's Process Safety Management standard. But, increasingly, there is a sense that good operating procedures are a vital part of running a smooth and safe operation. As plants become more complex, and the pressure to produce quality products with fewer shutdowns becomes more intense, there is less room for error. At the same time, many companies have lost a disproportionate number of their more experienced operators and supervisors during recent years, so the need for written instructions is greater than ever.

At first, writing operating procedures seems to be quite straightforward, particularly for plants that are already running. All that is needed is to write down how the plant is run, and to format the operating manual to meet regulatory requirements. However, the people responsible for this work soon find out that it is not so simple. Some of the problems that they run into include the following:

1. Extracting and organizing the information is quite difficult. The process includes reviewing original engineering documentation (which may be quite out of date), talking to operators (who do not always agree with one another), and observing actual operating conditions.

2. It is often very difficult to estimate the level of effort needed to complete the project. Usually, it winds up taking longer than anticipated and, what is often even more serious, it draws down the time of the experienced operators and supervisors more than expected.

3. The physical preparation and control of the final document can be more troublesome than anticipated.

4. It is often very difficult to obtain verified information concerning operating limits, i.e. the upper and lower bounds of safe operation. By definition, there is not much experience of operation outside

these bounds — otherwise there would have been far too many accidents. Yet these knowledge of these operating limits can be extremely valuable because they form the foundation of trouble-shooting guidance.

I have been working with operating procedures for a number of years now, and so felt that it would be appropriate to write down some of the ideas that I have had, and some of the lessons that I have learned (good and bad.) This book represents a summary of this experience. I have had the good fortune to work on a wide variety of projects, covering a wide range of industries such as petroleum refining, food processing and petrochemicals. Some of these projects involved writing manuals for grass-roots facilities, others were for plants that are already operating.

I would like to make a couple of editorial points. As the Disclaimer makes clear, all the examples that I use are fictional. Inevitably, many of the ideas are based on real-life projects; but I have made every effort to ensure that no company, process or individual can be identified. Second, I had to face the problem of inclusive language — the unfortunate fact is that the English language does not contain a neutral "he/she" word. Therefore, in the interests of space, and to avoid clumsy syntax, I have usually stayed with the masculine gender.

I would like to express his gratitude to all those people who have helped with the creation of this book. Val, Ann and Peter were as tolerant as ever. And many people have helped with invaluable critical reviews, particularly the following three people from Fluor Daniel: Dick Vandergriff, Joan Bradshaw and Duncan Smith. Naturally, any deficiencies or errors are entirely my responsibility.

This book is not complete, it never will be — we are all learning all the time. However, when I first worked in New York, I learned that marvelous phrase: "enough is enough, already!" So, in the words of Edmund Spenser, all that is left to say is: "Goe little book: thy selfe present."

CHAPTER 1
THE OPERATING MANUAL

INTRODUCTION

A well-written operating manual, one that is really useful to a plant's operators, will improve safety, profitability and environmental compliance for the following reasons:

1. The manual will encourage the development of consistent, authorized operations from person to person, and from shift to shift.

2. The process of writing the manual will stimulate discussions as to the best way of running the unit.

3. The writing of the manual will force management and the operators to define target conditions and acceptable operating ranges for the key variables.

In spite of these benefits, many plants do not possess a high quality manual. One reason for this is that it is possible to operate without one. Such an operation may not be optimal, but the plant will run. Consequently, the effort to get the procedures written is often given low priority.

Increasingly, however, this situation is changing. Companies are devoting more resources to writing and upgrading their operating manuals. In many cases, the immediate driving force behind this change comes from process safety regulations. But companies are also trying to achieve the genuine improvements in safety and operability that can occur when the operators have a high quality manual.

This book has been written to help those people in the process industries who are responsible for writing and updating operating procedures and manuals. Generally, these people are very knowledgeable about the process that they are designing, or the plant that they are operating, but they are not trained in how to develop, write or edit procedures. Therefore, the emphasis throughout this book is on the use of practical, proven techniques that allow high quality procedures to be written quickly and economically by people who have had little or no training in this area.

The specific goals of the book are as follows:

1. To show how to develop a design basis for writing operating procedures (the *Writer's Guide.)*

2. To show how a facility-wide standard can be developed that still permits reasonable flexibility from unit to unit and from operation to operation.

3. To describe how to set up and manage a procedures-writing project (the *Project Control Manual.*)

4. To describe how the operating manual can be published and distributed.

5. To show how to keep the manual up-to-date.

6. To discuss the requirements of the pertinent regulations, particularly OSHA's Process Safety Management regulation: 29 CFR 1910.119.

7. To provide guidance when evaluating or auditing an existing set of operating procedures.

8. To describe ways of doing all the above quickly and economically.

Because every plant is unique, there cannot be one single "correct" way of writing procedures that can be universally applied. For example:

- Procedures for batch processes will be different from those for continuous processes

- Procedures for the manufacture of pharmaceuticals will differ from those for refining oil

- Procedures written with strict government regulations in mind will probably look different from those where there is relatively little regulatory pressure

Furthermore, the goals of the procedures will differ from plant to plant within the same industry. Sometimes, the primary purpose of the procedures is to provide experienced operators with checklists for activities that they already know how to perform very well indeed. In other plants, the procedures will serve more as a training manual for new employees. Therefore, there cannot be a single "one-size-fits-all" operating manual. Instead management and the operators have to develop standards that are appropriate for their particular needs and circumstances.

For this reason, this book provides *guidance* for writing operating procedures, not *rules* or *instructions.* The book aims to stimulate thinking concerning the best ways of writing procedures for each particular plant and unit. So, for example, there is a discussion of the differences between *instructions* and *information,* and of the importance of keeping them separate. The conclusion reached here is that there should be separate volumes for each of these topics. Other people may reach a different conclusion depending on their own particular circumstances. There is also an emphasis in this book on the use of small, discrete modules for the operating procedures proper; some companies may prefer to follow a more discursive approach.

What really matters is that everyone go through the thinking process as to what methods and styles are best for their company and plant. No two situations are identical.

The above comments do not mean that the people in each unit can write procedures any way they choose, regardless of what their colleagues in other units are doing. It is important to set a standard that permits some flexibility. There are two reasons for this. The first is that it makes it easier for operators to transfer to another unit and to quickly learn how to operate it because they are familiar with the layout of the manuals. The second advantage of standardization is that it makes the writing process much easier. Instead of having to create instructions from nothing, the writers, to a certain extent, merely have to fill in the blanks. This reduces the problem of writer's block.

In addition to describing technical aspects of writing procedures, this book also discusses the management of a procedures-writing project. This includes issues such as scheduling, budgeting, the development of resource and publishing teams, the actual writing process, publishing and the review cycle.

Both the style of writing and the appearance of the final product are very important to the ultimate acceptance of the manual. In the words of the well-known proverb:

You never get a second chance to make a first impression

It is extremely important that the manual reads well and that it is attractive in appearance. The manual should be written in easy-to-read English, with plenty of graphics and other illustrations.

Organization Of This Book

There are four sections to this book.

The first section (Chapters 2 - 5) describes the structure and contents of the manual.

The second section (Chapter 6) describes the writing and publishing processes.

The third section (Chapter 7) discusses management issues, such as the selection of the procedures-writing team, scheduling and budgeting.

The final section (Chapter 8) discusses the integration of operating procedures with other elements of operations management particularly training, hazards analysis and the development of process information. This section also discusses process safety management regulations, particularly as they relate to operating procedures.

Throughout this book, a wide variety of examples is used. Although these examples are realistic, all the information in them is fictional, and has been created specifically for this book; none of it represents an actual plant or process.

Definition Of Operating Procedures

The definition of operating procedures used in this book is as follows:

Operating procedures are a set of written instructions that, when followed by the operations personnel, will minimize deviations from design or operating intent

The terms used in this definition are explained below.

Design Or Operating Intent

Before procedures can be written, management and the operators must clearly define how the plant is to be run; in other words they must determine the *design or operating intent* for their unit. This means that target conditions for all flows, pressures and temperatures have to be specified, along with the allowable deviations from those target conditions. When a plant is new, the target conditions are usually provided by the process engineering department — these represent the design intent. However, once a plant has been running for a few months it is likely that new operating targets will be set in order to increase efficiency or production. These new targets should be properly authorized through the plant's management of change system, and then incorporated into the operating procedures.

In practice, many plants do not have a clearly defined set of target conditions. In these cases, the development of these conditions is likely to be a significant part of the procedures-writing project. Indeed, problems with writing procedures are often a sign that the operating limits have not been properly defined.

A symptom of this can occur when there are different shifts of operators working on a particular unit. Each shift may have its own opinions as to target values and methods of operation. There is not usually any absolute right or wrong in such cases — just differences in judgment. Nevertheless, for each scenario, one mode of operation is necessarily safer and/or more efficient manner than the others. Writing the procedures brings these differences into the open and forces a resolution as to which set of conditions should be chosen. If handled properly, this can be a very exciting and rewarding process.

Design and operating intent must be quantified, wherever possible. For example, the following instruction:

- Increase the bottoms temperature gradually until the overhead pressure is about normal

can be re-written more precisely as:

- Increase the bottoms temperature by 5°C every 10 minutes
- When the overhead pressure reaches 3.4 kg/cm^{2}g stop the temperature increase
- The permissible overhead pressure range is 3.0 - 4.0 kg/cm^2

Written Instructions

Operating procedures must be written down in order to ensure that all of the necessary information has been communicated to the operators correctly. Sometimes, the operators may have a good set of verbal procedures that have never been written down. In these situations, the procedures-writing project consists largely of taking the verbal information that already exists, and writing it down in a clear and organized manner. The phrase "written down" does not necessarily mean that the instructions have to be on paper. Procedures can be recorded and communicated electronically.

Procedures must also be written down in order to satisfy audit and regulatory requirements. (Although it is not clear yet whether OSHA will accept electronically distributed manuals as meeting the requirements of the PSM standard.)

Operations

Operating procedures should be written by and for the operators and their line managers. There are other users, such as engineers and auditors, but generally their requirements are secondary; the operators are the primary customers of the procedures-writing project.

Different companies and plants use different words to describe the operations job functions that will be discussed throughout this book. The convention that will be used here is as follows:

1. An *Operator* is the person who actually carries out the instructions and guidance provided in the procedures. He or she often works on a 24 hour shift cycle. Many plants, particularly in the petrochemical business, refer to the people doing this job as *technicians*. Operators work both in the central control room and outside, in the field. Those plants that have implemented Distributed Control Systems (DCS) often put many of the control screens in one central facility, which may be a considerable distance from the plant itself.

2. The operator reports to a *Shift Supervisor*, who usually oversees the operation of a number of different units. Often, he is the person who is assigned full operational authority during an emergency.

3. The shift supervisor reports to the *Operations Superintendent*. This person, who does not work on a shift cycle, has broad management responsibility for running the plant. Above the superintendent there will be an *Operations Manager* and/or a *Plant Manager*.

Although the focus in this book is on *operating* procedures, many of the ideas and techniques can be used successfully in the development of *maintenance* procedures. The emphasis tends to be different however. For example, maintenance procedures will use more photographs and fewer flow diagrams. But other ideas, such as the use of modules, are very applicable. Maintenance Procedures are discussed further in Chapter 8 under the topic of Mechanical Integrity.

Operating Manuals/Operating Procedures

The phrases "operating manual" and "operating procedures" tend to be used interchangeably. Strictly speaking, an operating manual is made up of an integrated, coherent set of individual procedures, sometimes called SOPs or Standard Operating Procedures.

Some companies, however, use the phrase "operating manual" to describe "operating information" or "process safety information" (what is referred to as Volume II in Chapter 3 of this book.) They then refer to the actual operating instructions as SOPs.

Ownership

It is vital that the operators feel that they "own" the manual and that it reflects their knowledge, goals, concerns, opinions and needs. Ownership is always important, everyone likes to feel involved in what is going on and to feel that they are being listened to — this is basic human nature. In the case of an operating manual, if the operators feel involved, they will be much more willing to use it and to improve it than if it is merely provided to them by another group of people such as their technical staff or an outside consulting company. If it is accepted that the operators do indeed own the manual, then the task of everyone else (management, engineers, writers and consultants) is to provide assistance to the operators in the development of *their* manual — not the other way around.

The process of writing the procedures for the operators can raise a subtle, yet profound problem, which is that the way in which the procedures are written may not reflect the way in which the operators actually think and work. If the procedures are written in a rigid and highly structured manner, an experienced operator may feel that they do not really express how he actually runs the unit. Many operators tend to run their plant in a relatively eclectic manner — they perform numerous tasks in parallel depending on what "feels" right at that point in time. This is not inherently worse or better than the more formal or linear approach; it is just different. It is the responsibility of the people writing the procedures to make sure that not only are the procedures correct and well written, but also that they reflect the way in which the operators actually work and think.

ACCOUNTABILITY

One important reason for writing good procedures is to establish accountability. This does not mean that it is the purpose of a procedures-writing project to find ways of blaming the operators if something goes wrong, nor to provide them with excuses for thoughtless performance. Nevertheless, it is important that management and the operators know who is responsible for what actions, and to have a means whereby both parties can be held accountable for their actions.

SIGNS OF PROBLEMS

In order to quickly evaluate an existing set of procedures, it is suggested that they be perused for the following types of problems. If the problems listed here occur frequently, the procedures probably need to be re-designed and updated.

- Individual procedures are not dated or signed.
- The manual is incomplete.
- It is difficult to find instructions quickly.
- The manual is too bulky.
- The manual is difficult to use.
- Measurements, if provided at all, are given in instrument signals, such as divisions on a scale or "roots" of flow, instead of in engineering units such as °F or lb/h.

- Instructions are given in relative terms, such as "Open the valve two spokes."

- Excessive use of the words 'may', 'should', 'might' and 'ought.' Words such as these create ambiguity.

- Frequent use of words such as 'about' and 'approximately.' They indicate an undesirable lack of precision.

- Frequent use of words such as 'next' and 'then.' These words are generally redundant.

- More than one instruction per sentence.

- Operators are expected to make decisions based on non-objective and/or non-quantified criteria.

- Use of the passive tense. (Phrases such as "The valve will be opened" rather than "Open the valve".)

- Revision dates that are more than a year or two old. (Section (iv)(3) of the Process Safety Management (PSM) regulation requires that procedures be updated every twelve months.)

- Information mixed with instructions.

- The inclusion of hand-written notes.

- Excessive use of jargon, slang and abbreviations.

- Warnings and cautions that follow the instruction to which they pertain, rather than preceding it.

- Different formatting and layout styles within the same manual.

- The manual is physically damaged in some manner.

- The contents of the manual are not linked with the other elements of Process Safety Management.

JUSTIFICATION AND OBJECTIVES FOR A PROCEDURES-WRITING PROJECT

Although most managers will accept that writing a high quality set of procedures is a worthwhile activity, they also realize that their development can be expensive, and that the effort will tie up many of their key people for a considerable amount of time. It is therefore important to develop a justification for a procedures-writing project, and to clearly define the project's objectives.

SAFETY

It is estimated that between 40 and 70% of accidents are caused by operating error [1 2 3 4 5] (the phrase operat<u>ing</u> error is preferred to operat<u>or</u> error in order to remove the sense of fault-finding.) If the number of these errors is to be reduced, it is essential that the operators be provided with quality operating procedures, and that they are trained in their application.

ECONOMICS

The great majority of accidents do not result in safety problems, but they almost invariably lead to economic losses in the form of reduced production, lower productivity, damaged equipment and increased maintenance costs. A good operating manual should help reduce these losses by helping to achieve a smoother operation, one that has fewer upsets and shutdowns.

IMPROVED OPERATIONAL TECHNIQUES

In addition to reducing losses, the process of developing an operating manual may identify ways in which tasks can be modified or automated so that they can be conducted more efficiently or safely.

REGULATIONS AND STANDARDS

An operating manual is often required in order to meet regulations. (Although this is often the forcing function that gets the manual-writing project kicked off, it is desirable if the regulatory problem can eventually become almost an after-thought, and that the project justifies itself because it really does improve safety and operability.)

Many companies are writing or upgrading their operating procedures in order to meet regulatory requirements. Most process companies within the United States are required to comply with the OSHA Process Safety Management (PSM) standard, 29 CFR 1910.119, which has been in effect since 1992. This standard contains fourteen elements, one of which is concerned with operating procedures. The standard is described in detail in Chapter 8, particularly with respect to the writing of operating procedures.

Another regulation that requires companies within the United States to have quality operating procedures is the EPA's Risk Management Program (RMP). This has essentially the same intent as 1910.119, although its wording is somewhat different, and the number of affected facilities is substantially larger. The EPA is concerned with public health and safety, whereas OSHA focuses on worker safety. The RMP requires more documentation than the OSHA standard. At the time of writing, the final RMP regulation had not been issued, so only preliminary details regarding the writing of operating procedures are available.

Regardless of the details of the regulations, it is crucial to understand that the concepts behind them are inherently sensible (a statement which is not true of all regulations.) Therefore, even if a company does not *have* to address the regulation (for example, they may have sections of a factory which are not included in the standard, or the facility may be outside the United States), it may still be worth doing so in order to raise overall safety and operability standards.

Many companies are also improving the quality of their operating manuals in order to meet industry standards, such as ISO

9000 or the Responsible Care Program from the Chemical Manufacturers Association

TRAINING

Procedures provide the foundation for plant-specific training programs. Indeed, it is not possible to develop a complete training program until the procedures are written. Conversely, there is little point in developing procedures unless the operators are properly trained in their use.

Some companies refer to their operating procedures as a Training Manual. Generally, this is inappropriate — procedures are written for someone who is already trained. The Training Manual is another document.

TECHNICAL

Operating procedures can be used to satisfy other technical objectives. For example, the procedures will often include an information volume that provides the operators and other plant personnel with reference material such as Piping & Instrument Diagrams (P & IDs), Material Safety Data Sheets (MSDS), and Upper and Lower Operating Limits.

HUMAN ERROR

Earlier in this chapter it was pointed out that about half of industrial accidents can be attributed to human error. It is useful, therefore, to understand the nature of human error so that procedures can be written appropriately. In all cases, the distinction between error and fault must be kept in mind. If an operator makes an *error*, the *fault* may lie with the instruments or the equipment or the organization of the job. Human errors can be conveniently divided into one of three categories: Slips, Mistakes and Violations.

SLIPS

A Slip occurs when a person makes an error, even though that person knew what to do, and how to do it. It is formally defined here as follows:

A slip is a human error resulting from failure to carry out an intention, even though the person concerned had the capability, time, and equipment to successfully carry out that intention.

Slips usually occur during normal, routine, non-stress situations. For example, an operator may routinely take two samples from a certain section of the plant every shift, and he may have successfully performed this action hundreds of times. Then, on one occasion, he *slips* up and inadvertently switches the samples.

There is little that operating procedures can do to reduce slips because errors of this type are usually committed by persons who already know the operation well and so they are unlikely to be referring to the written manual while they are performing their work. The only exception to this may be through the use of checklists. A checklist, which is based on the written procedures, may help an experienced operator from "slipping up."

MISTAKES

A mistake (sometimes referred to as a cognitive error) occurs when a person acts on an incorrect train of reasoning, often because he was not properly informed as to what to do or how to do it. Mistakes also occur in the following situations:

- Someone incorrectly follows a set of unfamiliar instructions
- A person is inadequately trained
- When an unusual situation occurs and a new sequence of actions has to be devised, often under considerable pressure, such as during an emergency shutdown.

A mistake is formally defined here as:

A mistake is a human error that is a failure in diagnosis, decision making, or planning.

Mistakes can be further divided into errors that are "procedural" and errors that are "creative." A procedural mistake occurs when, for example, there is a lack of clarity in the operating instructions, thus causing the operator to misinterpret them. A creative mistake occurs when a brand-new situation presents itself, as sometimes occurs during an emergency, and the operator has to develop a response on the spot, often in a very short period of time.

Mistakes imply thinking, slips imply routine.

Procedures can be very effective at reducing mistakes because they provide guidance and instruction to the untrained operator, or to someone who is thinking through a new situation.

VIOLATIONS

Violations are a special type of error. They occur when supervision or management knowingly decides to over-ride the normal operating or safety procedures. This is not necessarily wrong. Indeed, managers are paid to exercise judgment and responsibility in abnormal situations. Sometimes, however, the decision to violate a standard results in an accident.

A common example of a violation is when operations supervision chooses to ignore a lab result or an instrument reading, either because they do not believe it or because they are prepared to run the plant regardless of the information that they have received.

If someone deliberately over-rides a standard operating practice, they need to write a temporary operating procedure that is good until normal procedures can be resumed.

TWELVE GUIDELINES

The various ideas and concepts that are developed throughout the remainder of this book can be summarized in the following twelve guidelines or aphorisms.

1. *The Operators Must Want To Use The Manual.* If the operators do not want to use the manual, they will not use it. It cannot be forced on them. Therefore, it is the responsibility of the people writing the manual to make sure that they develop a product that the operators find really useful.

2. *The Manual Must Be Useful and Useable.* Many manuals do not really address the operators' needs. For example, many manuals provide too much process information, but not sufficient help with operations and troubleshooting, even though this is what most operators want. Furthermore, many manuals are physically difficult to use. They should be of a convenient size, easy to read, colorful, and should contain plenty of graphics.

3. *Although The Manual Looks Like A Book, Write It Like A Data Base.* Although the manual is a "book" in the physical sense of the word, it is much closer to being a data base in the way that it is used. Therefore it should be "written" like data base code.

4. *Distinguish Between Procedures, Information, Guidance And Training.* Of all the concepts discussed in this book, this may be the one that is the most important. Most operating manuals are designed to meet multiple goals, such as training new operators, providing checklists for experienced operators, and meeting regulatory requirements. There is nothing wrong with having multiple goals — *as long as they are carefully distinguished from one another in the manual.*

5. *Develop A Structure Of Publicly Declared — But Privately Maintained — Modules.* Often, the biggest practical problem with operating manuals is not the writing of the first edition, but keeping them up to date. The concept of modular construction,

that was developed for software, and that is explained in the next chapter, is one way of addressing this problem.

6. *Design The Manual To Be Complete — Design The Manual To Be Incomplete.* The manual should always be an integrated, holistic document from the very beginning of the project. At first, there will be almost no detail — this will be added as the project progresses. But there will always be a complete structure, to which more and more detail can be added.

7. *Use Graphics and Text.* Many manuals contain very little graphical material. This makes them unattractive and difficult to use. Graphics can include P & IDs, sketches and photographs.

8. *Use Two-Page Modules Because That Is What The Reader Sees Anyway.* When an operator opens the manual, what he actually sees are two pages. It is suggested that each procedure be designed to fit these two pages.

9. *Make All Modules Consistent.* It is important that the procedures be as consistent as possible from unit to unit and operation to operation. This does not mean that they have to be identical — but there should be an overall similarity. Not only does consistency help the operators as they move to new units, it also greatly facilitates the writing process by reducing writer's block.

10. *Write Minimally.* Minimalist writing is writing that eliminates all extraneous material. It is not quite the same as writing with just a few words — if many words are needed then they should be used. But every word must actually help the operator run his unit. All extraneous materials should be ruthlessly purged.

11. *Make Distribution Of The Manual Interactive.* The manual should not be written by outsiders and then "given" to the operators. Instead, the final users should be the ones who direct what it should look like, and what it contains.

12. *It's A Project — Treat It Like One.* A common problem with the writing of procedures is that the activity is not treated as a project

in the same way as say installing a new piece of equipment is handled. This can lead to serious problems. Developing and writing the manual should incorporate all normal project controls, including budgets, schedules and progress reports.

The Users

Although operating procedures are written primarily for the plant operators, there will be other users who need to be identified at the start of a procedures-writing project so that the manual reflects their needs appropriately.

Experienced Operators Familiar With The Unit

Experienced operators and supervisors need procedures that are short and easy to use; they do not normally need a detailed description of how to run a process that, in many cases, they have worked on for years. Generally, they are looking for checklists and help with troubleshooting. Checklists are useful when they are working through a long sequence of actions. No matter how familiar these actions may be, a checklist will help ensure that a step does not get overlooked, or that two actions are not carried out in the wrong sequence. In other words, checklists can reduce slips.

Troubleshooting assistance is particularly useful to experienced operators because it is the one area in which they can never have sufficient knowledge.

Experienced Operators Unfamiliar With The Unit

If an operator is transferred from another section of the facility, he will need unit-specific training for the new process. It is assumed that this operator will be fully conversant with the procedures for operating process plants in general, and that he knows and understands company policies and general practices. The operating procedures will provide the plant-specific details that he needs.

INEXPERIENCED OPERATORS

Operators new to process plants need basic training. This training will be supplemented with procedures from actual units so that they can gain experience in the various plants. In spite of this use for procedures, the procedures themselves should not generally be written at this basic level; instead they should be written for someone who already has been trained.

TECHNICAL STAFF/MANAGEMENT

The technical staff use the procedures as a framework within which they can conduct plant tests and assist with troubleshooting. Management may also use the procedures when planning large projects such as a turnaround or revamp.

TRANSLATORS

Many manuals are written by an engineering company for use by a client in a different country. This means that the first "user" of the manual will be a translator. If this is the case, it particularly important to avoid jargon, ambiguity and complex sentence structures.

AUDITORS AND INSPECTORS

Auditors and inspectors may be either employees who are working directly for the plant, or they may be from an external body such as OSHA or the company's corporate office. Auditors are likely to evaluate operating procedures during an overall safety review because procedures are easy for an outsider to follow and understand. It is relatively simple to check on their completeness, clarity and whether they correspond to what the operators are actually doing in the field. This is in contrast to other elements of a safety management program, such as PHAs or Employee Participation, which are generally much more difficult to inspect and validate, and which cannot therefore be audited as easily.

One reason for making operating procedures attractive to look at is that it will make a good impression on auditors; it will quickly show them that care and expense were devoted to this element of the Process Safety Management program.

DIFFERENT INDUSTRIES

The way in which a set of operating procedures are written will depend on the industry for which they are being written.

THE PROCESS INDUSTRIES

The process industries fall under SIC (Standard Industrial Classification) Codes 28 and 29: *Chemicals and Allied Products, and Petroleum Refining and Related Industries*. Included within this are chemical plants, petrochemical plants, refineries, oil and gas production facilities, food-processing, and pharmaceuticals. These industries are not homogenous; some of the differences between them are reviewed below.

Refining

Petroleum refineries are flexible in the way in which they operate and in the way in which they can structure their processes. They work with constantly changing feedstocks and product slates. Partial turnarounds are quite common, and they can often handle multiple operations with the same equipment.

Petrochemicals

Most petrochemical plants are less flexible than refineries. They have a single train of equipment making a limited number of products. Heavy emphasis is often placed on simply keeping the plant running. Emergency response often has to be very quick, for example in the case of a runaway chemical reaction. Therefore, there may be a high dependency on the automatic instrumentation and interlocks.

Many of these plants have adopted ISO 9000, particularly if they sell to European customers.

Oil and Gas Production

Theses processes tend to be quite simple compared to refineries; consequently there is often very little operator presence, indeed the plant may be completely unattended at times.

Specialty Chemicals

These plants tend to be batch-oriented which carry out multiple activities with the same items of equipment.

FOOD PROCESSING

Operating conditions in food processing plants are usually relatively benign and they do not normally handle large quantities of hazardous chemicals. However, the standards of health and hygiene require very close monitoring of the operations. The processes tend to be batch oriented. There are many regulatory requirements from agencies such as the Food and Drug Administration.

THE NUCLEAR POWER INDUSTRY

A pivotal industrial accident in the United States occurred in 1979 at the Three Mile Island Power Plant in Pennsylvania. The accident is often referred to by the initial letters of the plant: TMI. The severity of the accident was magnified due to a mis interpretation by management and the operators of the information that they were receiving from the plant's instruments. They "fixated" on instruments that were providing bad information, even though many other instruments were correctly indicating what had gone wrong.

These misunderstandings were exacerbated, in part, by limitations in the design of the operator-machine interface. Partly as a result of TMI, the nuclear power industry in the United States has invested heavily in the development and understanding of operating procedures, and in many other aspects of human factors analysis.

The process industries and the nuclear power industry share considerable technical similarities. For example, both transfer large quantities of gas and liquid from one place to another as a means of transferring energy. Yet, in spite of the technical similarities between the two industries, there are also important differences, some of which are listed below.

- The nuclear power industry is publicly funded via regulated utilities. Therefore, it is not under the same explicit financial pressures as are commercial process companies. This means that the nuclear power industry is more able to provide funding for extensive studies and test programs.

- The two industries tend to operate to different schedules. Generally, commercial companies are under greater pressure to complete projects more quickly and to return money on any investment more quickly.

- A nuclear power plant is inherently simpler than most process plants (although it does have very sophisticated safety instrumentation.) Basically, it is a boiler and generator. Most process plants are considerably more complex than this. Typically, they handle a far wider range of chemicals, many of which are very hazardous, and there is usually a much greater variety of processing steps.

- The public is extremely sensitive to even the most minor accidents at nuclear power plants due to a deep-seated fear and distrust of radioactivity. In spite of increased environmental concerns, process plants do not have to cope with such a high level of sensitivity because their worst-case accident is not so serious (as discussed in Chapter 5.)

- There is more standardization among nuclear power plants than among process plants. In some countries, such as France, duplication of plants is a deliberate policy. One of the advantages of doing this is that the cost of developing and writing procedures can be amortized over a large number of facilities.

- Because nuclear power is tightly controlled by government agencies, everything that they do, including operating procedures, tends to involve a lot of paper work.

In summary, there are many useful lessons to be picked up from the nuclear power business, but the differences in culture make for very different approaches.

MULTIPLE GOALS

Operating manuals are usually expected to serve different goals. There is nothing inherently wrong with this, as long as the nature of the separate goals are understood and the manual is organized so that they are handled appropriately.

The following is a list of some of the goals that a manual is often expected to meet.

1. Provide brief, checklist-type instructions to operators who are fully trained and who are very familiar with the plant and its operation.

2. Provide training for operators who are experienced in general plant operations, but who do not know the particular unit.

3. Provide background and reference information, such as Material Safety Data Sheets (MSDS), limits of operating conditions and telephone numbers of vendors.

4. Describe what to do in the event of a major emergency.

5. Describe what to do in the event of an environmental upset. This could include instructions as to how to contact the appropriate regulatory agencies.

Some examples of the confusion of purpose that is often found in manuals when these goals are not clearly distinguished are shown in Tables 1 and 2.

Table 1

Example of Confusion of Purpose (1)

Mixing Instructions with Information

PROCEDURE FOR FILLING THE WEIGH TANK	
Step 1	Start the Dilution Oil Pump, P-1205
Step 2	Add ten (10) 50 lb bags of catalyst powder to the Weigh Tank, T-1230.
Step 3	Add 200 gallons of process water to T-1230 using batch meter FQ-1209. This will dissolve the catalyst powder.
Step 4	Measure the density of the solution. It should be in the range 1.105 - 1.115. If it is not, go to *Step 10.*
Step 5	Add an additional 100 gallons of water

The second sentence in Step 3 — "This will dissolve the catalyst powder" — is redundant. It is not an operating instruction and therefore, strictly speaking, should not be included in the "*Procedure* for Filling the Weigh Tank." It is actually an item of process information that should be located elsewhere. Although a single instance such as this is trivial, its repeated occurrence will materially degrade the usability of the manual.

Table 2

Example of Confusion of Purpose (2)

Mixing Instructions With Training

PROCEDURE FOR FILLING THE WEIGH TANK	
Step 1	Start the Dilution Oil Pump, P-1205. Do this by pressing the green “ON” switch at the pump.
Step 2	Add ten (10) 50 lb bags of catalyst powder to the Weigh Tank, T-1230.
Step 3	etc. . . . .

The second sentence in Step 1 of Table 2 (“Do this by . . . ”) would normally be part of the training program, not the operating procedures. Once more, in isolation this problem is trivial. But if it is repeated throughout the manual it can become serious.

Almost all operating manuals suffer from the problem of confusion of purpose to some degree. One side benefit of exerting strict discipline in this area is that the operating procedures proper, i.e. the manual that contains just the actual operating instructions, can become surprisingly short.

FIVE VOLUME STRUCTURE

In order to minimize confusion of purpose, it is suggested that the operating procedures be organized into five volumes. These volumes do not have to be in separate binders, but they do represent a convenient way of organizing the operating procedures. They are:

Volume I — *Operating Instructions*
Volume II — *Operating Information*
Volume III — *Emergency Operating and Shutdown Procedures*
Volume IV — *Troubleshooting Manual*
Volume V — *Initial Start-Up Manual* (where needed)

VOLUME I — OPERATING INSTRUCTIONS

Volume I is the heart of the manual because it tells the operators how to run their unit. It is what most people think of when they hear the term "Operating Procedures." It in turn can be divided into the following sections:

Section 1 — Startup
Section 2 — Shutdown
Section 3 — Normal Operations
Section 4 — Temporary Operations
Section 5 — Generic Procedures
Section 6 — Vendor Manuals

A simple way to decide whether a sentence belongs in Volume I is to determine if it is structured in the form of an instruction. Sentences such as the following belong in Volume I because they are all "do" items.

- Open the valve
- Measure the temperature
- Inform the supervisor

The following statements, on the other hand, belong elsewhere in the manual.

- The valve is made of 316 stainless steel
- The maximum allowable temperature is 165°C
- The fire department telephone number is: xxx-yyyy
- The vibration may be caused by a plugged filter

These statements are either information or troubleshooting guidance — not operating instructions. Therefore they do not belong in Volume I.

Volume II — Operating Information

Volume II contains information that the operators may need to refer to on an as-needed basis. Examples include:

- Safe work policies and procedures
- Operating limits, such as maximum and minimum values for temperatures and pressures
- Descriptions of process chemistry
- Engineering data

This information is particularly valuable when the unit is being started up for the first time. Therefore, it should be linked closely with Volume V, *The Initial Start-Up Manual.* Volume II will address many of the Process Safety Information (PSI) requirements of OSHA's PSM standard (see Chapter 8.)

Volume III — Emergency Operating and Shutdown Procedures

Most plants have two types of emergency procedures. The first is for the entire facility and provides instructions on overall emergency response. It is often referred to as *The Emergency Response Manual.* It provides information on issues such as fire control, the mitigation of toxic gas releases and coordination with external agencies.

In addition, each operating unit will have its own set of process-specific *Emergency Operating and Shutdown Procedures,* referred to here as *Volume III.* These procedures describe how to recognize and handle an emergency on that particular unit.

VOLUME IV — TROUBLESHOOTING MANUAL

Troubleshooting is different from emergency response. In an emergency situation action must be taken quickly — there is no time for thought or discussion, and the results of failing to take the correct action could lead to a catastrophic accident (or an existing catastrophe will become worse.) Troubleshooting, on the other hand, addresses non-crisis situations, usually to do with product quality or equipment reliability. Generally, there is plenty of time to think through these situations, and there is usually a range of feasible responses. The *Troubleshooting Manual*, Volume IV, provides operators and their management with ideas and suggestions as to how such situations can be handled.

VOLUME V — INITIAL START-UP MANUAL

Volume V, *The Initial Start-Up Manual,* is written when the plant is brand new and there is no plant-specific experience on which to draw. It can also be used after a major turnaround. It covers mechanical completion, precommissioning and commissioning.

WRITER'S GUIDE

The Writer's Guide provides assistance with the actual writing and formatting of operating procedures. One of the goals of this book is to provide sufficient guidance to the reader such that a Writer's Guide can be prepared.

PROJECT MANUAL

A common problem with the writing of procedures is that the activity is not treated as a project in the same way as say the installation of a new piece of equipment. Instead, the writing of procedures is something that is "fit in" with other work. Not surprisingly, this leads to delays and product-quality problems. The writing of procedures should be treated as a project, with its own schedule, project staff and budget, in just the same manner as a capital expansion project. This means that there should be a Project Manual for writing procedures. This manual describes staffing, financial and scheduling issues.

Maintenance of the Procedures

Once the procedures have been written and published, they have to be kept up to date using a Management of Change system. Even if the first version of the procedures is well-written, maintaining them is often a problem. In almost all plants, equipment and operating conditions are constantly changing. Many of these changes will create a need to update the operating procedures because the procedures describe the interface between the operators and the plant. Therefore many, indeed most, changes will lead to changes in that interface. These include:

- Modifications to equipment
- Changes in instrument set-points
- Changes in the company organization

The need for a quick, efficient and easy system for updating procedures is a major justification for writing them in a structured, modular manner, as described in the next chapter.

A related issue concerns the handling and distribution of the updates. Often, there are multiple copies of the manuals in circulation. Many of these may be temporarily lost, misplaced or buried in lockers. Also, many of the manuals will have "walked." Ensuring that the new pages are inserted and the old pages removed whenever a change is instituted is a challenging task, and provides a powerful incentive for using electronic distribution.

Comparison With Software

Many useful analogies can be made between software and operating procedures. Both businesses are concerned with organizing a large amount of disparate information into a structured, compact, easy-to-use package that can be easily maintained. Therefore, it is beneficial to review the lessons learned by the software business and to apply them where appropriate, to procedures.

Five areas have been selected here for discussion. They include:

1. Data Base Comparison
2. Desk-Top Publishing
3. Top-Down Programming/Writing
4. Expert Systems
5. Project Management

DATA BASE COMPARISON

The language used when talking about operating procedures is that of the world of books. An operating manual looks like a book because it has words printed on pages and because it is bound in a binder. As with a book, the process of creating the manual is referred to as "writing" and the operators are said to "read" it. Yet these semantics are misleading. Virtually all books progress sequentially — that is, there is a narrative with a beginning, a middle and an end. Not only is this true of novels, mystery stories and biographies, it is also true of technical books, such as this one. Typically, they start with an introductory chapter, which is then developed in greater detail in the subsequent chapters. Then, at the end of the book there is often a review and a conclusion.

Operating procedures are not written in a narrative form because they are not read straight through as if it were a book. Instead, procedures are much closer in form to random access data bases [6] because the operators use the information in them randomly, i.e. they go to the section that addresses their needs, regardless of its location within the manual. Moreover, one piece of operating information, such as starting a particular pump, may be used in more than one operating sequence. This feature, the multiple use of one set of information, is fundamental to data base design.

Large Amount of Information/Indexing

Operating procedures usually contain a lot of information, just like computer data bases. They have to be designed so that the information can be stored, retrieved and updated in an orderly manner. If an operating manual is viewed as being a data base, rather than a book, then it is very important that a good index be provided because the reader will need to be able to find information quickly and easily.

Randomness

Information in a set of procedures will always be random, at least to some degree. During the development of the first version of the procedures, the operating modules may correspond to the order in which they are normally used. However, once the procedures have been modified and updated, the information will be stored in the manual in an increasingly random manner, just as it is in a data base.

Similar Structure

Individual operating procedures should have a standard structure and format because it makes moving around the manual and finding the relevant information so much easier. Once more, this is very similar to the way in which data bases are organized.

Modular Structure

Large computer programs consist of hundreds of separate modules (often referred to as sub-routines or procedures) that carry out a function such as adding two numbers or drawing a line on a screen. These modules call one another via their title blocks. This makes it possible to upgrade or maintain one particular module without worrying about the effects on the rest of the system.

Failure to follow a structured, modular design leads to what is known as “spaghetti code.” When this happens, the program eventually becomes too cumbersome and complex to be maintained because it is almost impossible to figure out what is going on internally.

DESK-TOP PUBLISHING SOFTWARE

Desk-top publishing software is used to develop presentation quality printed materials. It is widely used in the preparation of newsletters and formal reports and is very useful when writing an Operating Manual. The software takes text and graphics from other programs and sources. They can then be laid out so as to look as attractive as possible. Many modern word processing programs include desk-top publishing features.

TOP-DOWN PROGRAMMING/WRITING

The basic idea here is that the manual will *always* be a complete structure, no matter what stage the project is at, and no matter how little detail may have been developed. In the early stages of a procedures-writing project, there will be very little detail, but the overall structure will always be in place from the very beginning. Then, as the project proceeds, more and more detail can be added to this structure.

EXPERT SYSTEMS

An expert system is a computer program that draws on a data base of experience, in this case practical plant experience. It can be used for assistance with troubleshooting. It is built up by asking experienced operators how they would respond to issues such as product quality problems, excessive consumption of utilities, or incipient mechanical problems. Generally, there is no one single right answer to questions of this type. Each person will have an opinion; and most of the opinions will have validity. The person writing the expert system will organize all these opinions and ideas and enter them into the data base. Then, at a later date, a person with less experience can call on this program to help them solve problems that they do not know how to tackle.

Volume IV — the Troubleshooting Manual — is essentially an expert system that is written down on paper.

Project Management

The final comparison that can be made between software and operating procedures concerns the management of their development. The software industry has developed many techniques to ensure that their programs come in on schedule, on budget and with as few as errors as possible. These techniques can be applied to the writing of procedures.

Auditing The Procedures

It may seem strange to be discussing the audit process in the first chapter of this book, before methods for writing procedures have even been introduced. Yet no management system is complete unless there is a systematic method for checking that all the steps are in place and that they are being properly used. (The word “audit” has unfortunate connotations because it is often associated with legally-based proceedings, such as those conducted by the Internal Revenue Service. Failure to meet the requirements of the law in these cases usually means that a penalty will be imposed. For this reason, the audit may be referred to as a “review” or “survey” — but it is still an audit.)

External Audits

External audits may be conducted by an outside agency, such as OSHA or by another group from within the company. An auditor must work with some objective, independent standard. He or she must also have the authority to take action if problems are uncovered. It is essential that the auditor be independent for the following reasons.

- If deficiencies are found, the auditor should be able to report them without fear of retribution. Therefore, the auditor should always report to someone who is outside the plant’s normal chain of command.

- His independence will enable him to look at situations objectively and with fresh eyes. Even in even the best run plants, people learn to live with situations that outsiders immediately identify as requiring corrective action.

- He will have experience in other plants, and possibly other companies. This provides an invaluable background against which to evaluate the present situation.

Internal Audits

Most audits or reviews will be internal. Indeed, the operators and supervisors should be continually evaluating the system and their own performance. For example, in some Japanese companies, if an operator wishes to change the position of a valve or switch, he will first point to that item and speak out loud what he plans to do, and why. Only after doing this will he then take the action. This may seem a little bizarre, but it does constitute a quick and effective audit. Other companies have a "take two" program which achieves the same goal. The operator is encouraged to take a couple of seconds to think through the consequences of his action before he actually takes that action.

The concept of "Action — Response" instructions, that is discussed in the next chapter, provides an excellent form of an internal audit.

CHAPTER 2
VOLUME I — THE OPERATING PROCEDURES

INTRODUCTION

In Chapter 1 it was suggested that an operating manual be developed as a series of five volumes. If this approach is followed, the operating procedures proper will be located in Volume I. This chapter provides suggestions as to how Volume I can be organized and developed. Typically, it will contain six sections:

Section 1 — Start-Up
Section 2 — Shutdown (excluding emergency shutdowns)
Section 3 — Normal Operations
Section 4 — Temporary Operations
Section 5 — Generic Procedures
Section 6 — Vendor Manuals

Each section is composed of modules that provide detail on how to conduct specific operations, and that show the linkage between the modules.

MODULAR DESIGN

MODULAR STRUCTURE

On most plants, change is a constant. Almost every day there is a change to the equipment, the organizational structure or the operating targets. Most of these changes require that the operating procedures be updated because the procedures represent the interface between the plant and the people who are running it. Unfortunately, most operating procedures do not readily lend themselves to being updated because of difficulties with the system-wide impact of apparently small changes. A small modification to a procedure in one place may lead to changes elsewhere in the manual. This in turn leads to a major re-formatting effort. Even the chore of simply identifying where the changes show up can be formidable.

Partly because of these problems, most companies do not update their operating procedures as frequently as they should. Therefore, after a few months the manual is out-of-date; after a year or two it is often so badly out-of-date as to be unusable or untrustworthy.

For these reasons, it is very important to design the procedures for ease of maintenance and updating. In the long-term, this is likely to be more important than expediting the production of the first edition of the manual.

As has been discussed in Chapter 1, another business which has struggled with the same issues is the software industry. Modern computer programs are extremely large, and the effects of even small changes in the programs can be both far-reaching and subtle. In response to this, software engineers make strict use of the concept of "publicly declared modules that are privately maintained." What this means is that all software code is divided into manageable modules. Depending on the computer language being used, these modules are called sub-routines, functions or procedures. Each module has an explicitly defined purpose and function, and can be "called" by any other module. Modules can only call one another through their title blocks. It is not possible to move from one module to the heart of a second module.

The advantage of this approach is that, if a module has to be changed, its internals can be updated and modified "privately maintained" — without having any impact on the overall structure. As long as the title of a module is not changed, a writer is free to change the contents of a module without having to worry about system-wide implications. (As discussed later, there will be some exceptions to this assumption.)

As a very simple example of this, consider the software schematic shown in Figure 1. On the left is a sequence of computer instructions — the flow of the program is from top to bottom. One of the steps requires that two numbers, X and Y, be added to one another to give the answer Z. To do this, the main program calls a subroutine with the title ADD. The numbers X and Y are fed as arguments to the subroutine. It adds them and then returns the answer Z to the calling program.

Figure 1

Computer Subroutine

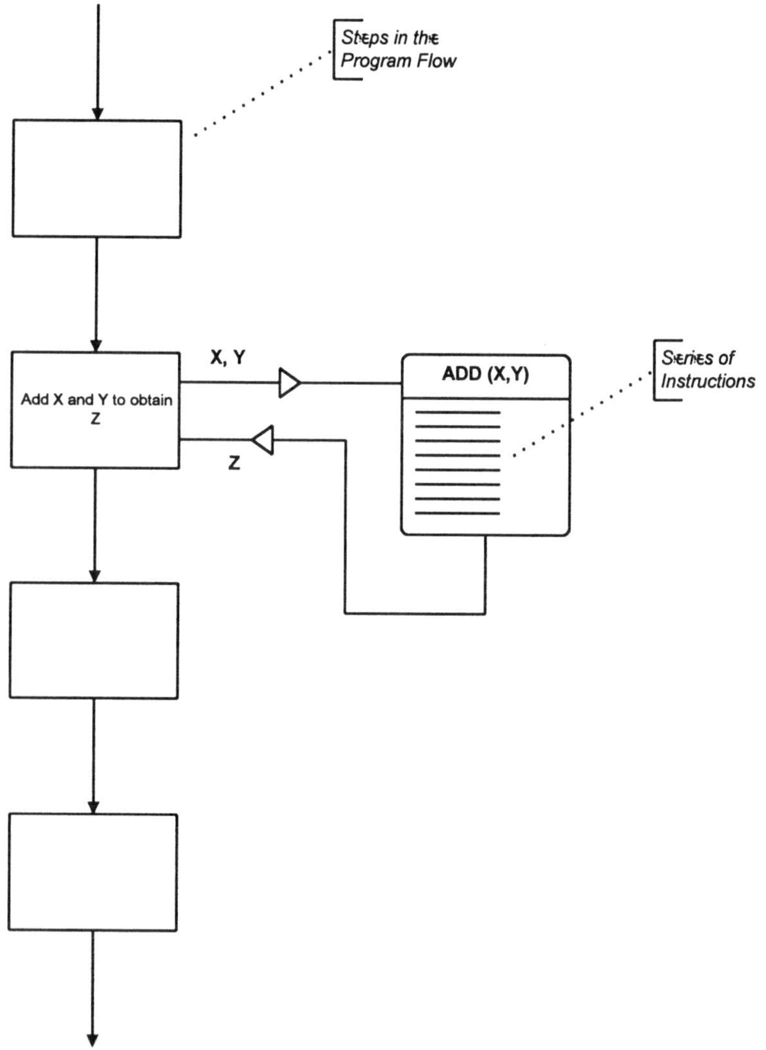

If, after this program has been issued, it is decided to change the module ADD, it can be pulled out of the system, modified and reinserted without having any effect on any other part of the program (although system effects do need to be checked for.)

Exactly the same principle can be applied to operating procedures. Figure 2 shows a series of operating instructions, analogous to the main program in Figure 1. Each of these instructions is the title of an operating module.

The detailed module, which is analogous to a software subroutine, for "Heat V-100" is also shown in Figure 2.

Figure 2

Operating Instructions

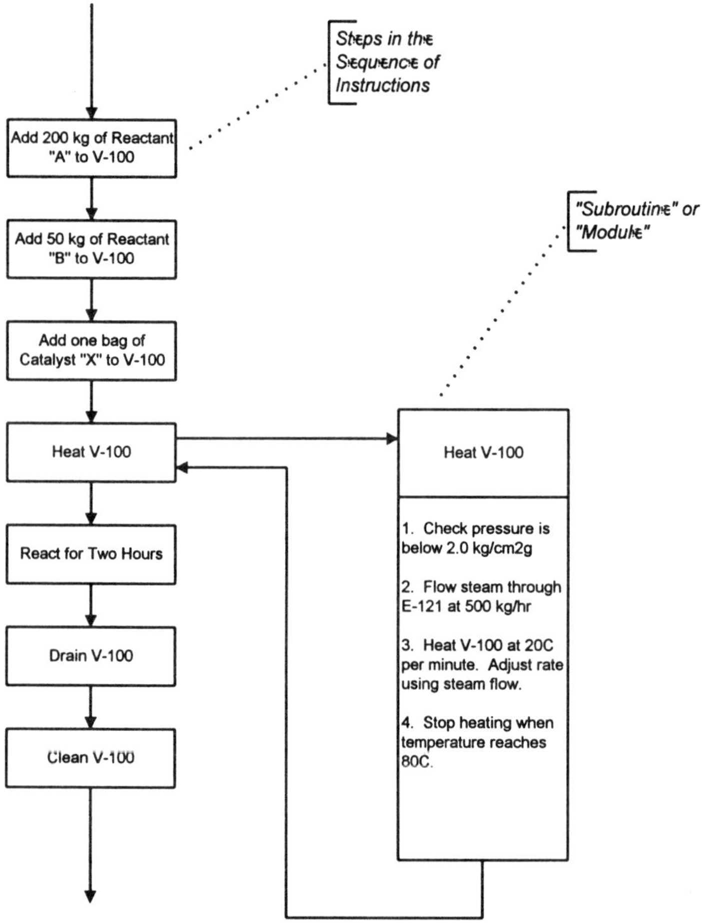

One of the rules of this process is that a module can only be entered via its title block. So, the sequence shown in Figure 3 is not allowed. Instead of going through the title, the program has skipped straight to one of the operating instructions within a module. This is not permitted.

Figure 3

Incorrect Use of Modules

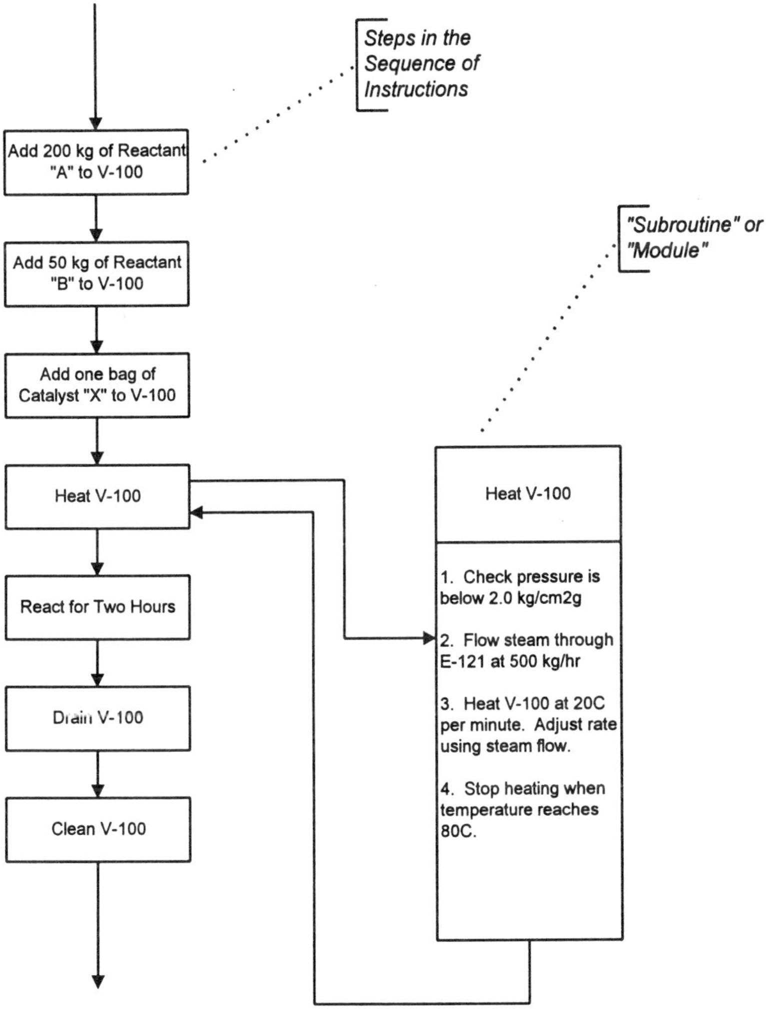

The advantage of using just title blocks to access modules is that it makes maintenance of the procedures much easier. For example, a pump named P-100 may have three modules associated with it:

- "How to start P-100"
- "How to stop P-100", and
- "Emergency Shut Down of P-100"

If, after the manual has been issued, it is decided to change the pump's driver from an electric motor to a steam turbine, all that the procedures-writer has to do is to pull these three modules out of the manual, rewrite them to reflect changes in the operations, and then re-insert them into the manual. The only way other modules in the procedures can access instructions about P-100 is through one of these three titles. Since the titles themselves have not changed, the person responsible for updating the manual with the new information concerning the operation of P-100 does not have to hunt through every page of the manual to find out whether there are any additional instructions that require modification.

MODULE LINKAGE

The linkage of the modules may involve some branching based on conditional instructions, as seen in Figure 4, which shows a sequence of four activities in series. They are:

Module 1 — Start Gasoline Flow Using Pump P-100
Module 2 — Fill V-100, OR
Module 3 — Fill V-101
Module 4 — Stop P-100

There is a conditional branching statement in the first module.

Figure 4

Sequence of Modules

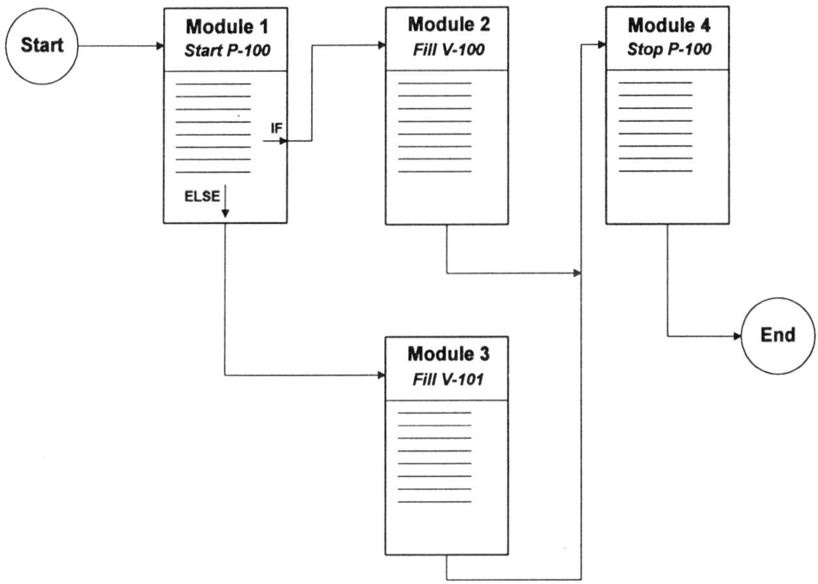

Figure 4 demonstrates that, although it is only permissible to enter a module through its title block, it is possible to have more than onc cxit from it.

In general, each module should be restricted to about ten instructions or less, and should not extend for more than two pages of text. If a module becomes bigger than this, it probably should be broken into two. So, for example, if the module with the title *Start-up of Product Stripper Feed Pump, P-2013* becomes too long, it can be broken into:

- *Lining Up the Product Stripper Feed Pump, P-2013*, followed by
- *Starting the Product Stripper Feed Pump, P-2103*

TOP-DOWN DEVELOPMENT

Top-Down Structure

For a large operating unit, it is useful to follow a top-down approach in which the instructions are developed in increasing levels of detail. Consider the example shown in Figure 5. There are five levels of increasing development.

- The first level is simply a single block — *The Operating Manual.*

- One level down the unit is divided into four major operating areas: Utilities and Units 100-300. The division into units can be made by dividing the plant into operating sections, where each section can be isolated from other sections and where each section can be operated more or less independently of the others.

- At the third level, each of the operating areas is divided into five types of operation as shown (in Figure 5, this is shown for Unit 100 only.)

- Each of the operating areas is further divided into operating blocks. Block 3: *Start Compressors*, is shown.

- At the fifth, and final level, are the actual modules. There are four of them for this particular block: Modules 3-1 to 3-4.

Figure 5

Top-Down Development

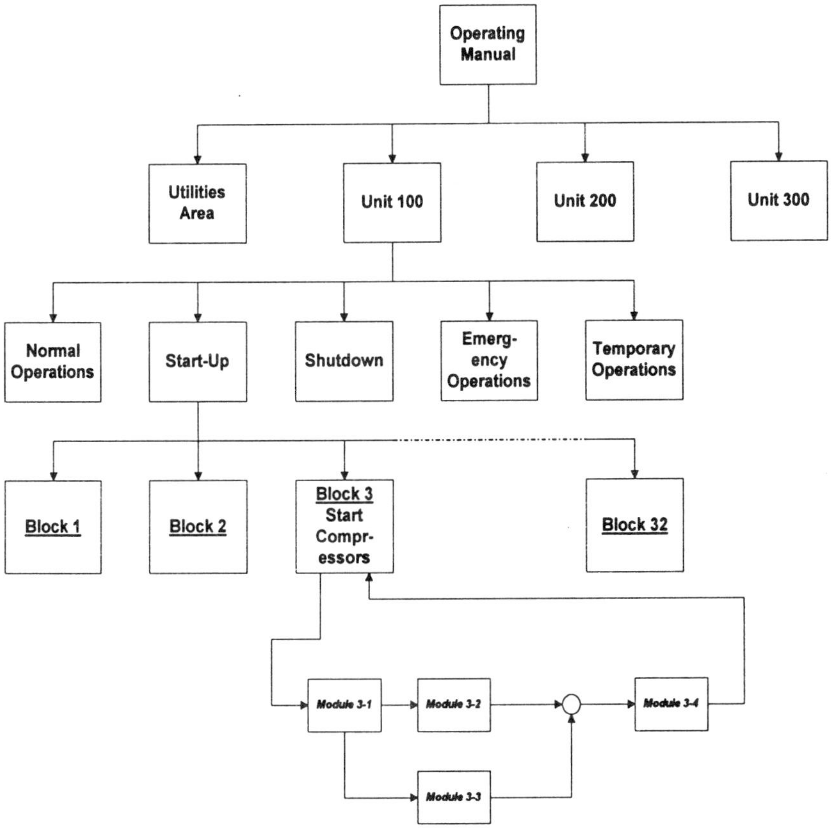

This approach ensures that, all times, the operating manual is always complete, at least structurally. Naturally, at the start of the project there is very little detail. But, detail is added *within the pre-designed framework*, hence the scope is always defined. This approach does not mean that the structure cannot be changed. Just the opposite is the case. As detail is added, it will become apparent that the structure will have to be modified, expanded or contracted. For example, it may be necessary to divide Block 3 in Figure 5 into two separate blocks. There is no problem with doing so.

Linkage

Figure 5 shows how the modules are developed in a top-down manner. It also shows how the modules link to one another. For example, Modules 3-2 and 3-3 are carried out in parallel, and both must be completed before Module 3-4 can be started (this concept was also illustrated in Figure 4.)

In principle, the same approach can be used for the Blocks that are one level above the modules. There is not necessarily a requirement that Blocks 1, 2 and 3 have to be carried out in the order in which they are numbered.

It is also possible to have generic modules, with titles such as "How To Catch A Sample Of Sour Water" that are essentially free-standing. They can be called upon by other modules on an as-needed basis.

Module Design

Operating procedures should be written so as to be consistent with one another. Each module should have exactly the same appearance as all the other modules. There are two reasons for this.

1. A consistent style makes the operator's job easier. If the procedure modules are all the same design, the operators will find it much easier to locate the information that they need, even if the procedure itself is unfamiliar to the operator.

2. Consistency helps the writers. By pre-designing the format as much as possible, the writers are less prone to the problems associated with writer's block. Most people find it much easier to "fill in the blanks" than to write creatively on a plain piece of paper.

It is suggested that each module be divided into three principal sections:

1. The Title Block
2. The Operating Instructions
3. The Authorization Block

They can be arranged as shown in Figure 6. The location of the boxes on the page is not critical. For example, some companies prefer to place the authorization block at the top of the page.

Figure 6

Three Section Module Layout

<table>
<tr><td>Title Block</td><td></td></tr>
<tr><td colspan="2">Operating Instructions</td></tr>
<tr><td>Step 1</td><td></td></tr>
<tr><td>Step 2</td><td></td></tr>
<tr><td>Step 3</td><td></td></tr>
<tr><td>etc.</td><td></td></tr>
<tr><td></td><td></td></tr>
<tr><td>Authorization Block</td><td></td></tr>
</table>

Almost all procedures have this three part structure, although it is rarely called out so explicitly.

The Title Block

The Title Block, an example of which is illustrated in Figure 7, contains the following elements. (In practice, it will not usually be necessary to include them all, but they are shown here for completeness.)

Figure 7

Example of a Title Block

MODULE NAME	*Starting P-100, The Recycle Charge Pump to the Lights Column, C-121*	
PURPOSE OF THIS MODULE	This module describes how to transfer isobutane and other light hydrocarbons containing H_2S to the Light Ends Column, C-121	
SPECIAL SAFETY CONSIDERATIONS	**Normal precautions regarding H_2S and LPGs.**	
DISCUSSION	V-100 sight glass is difficult to read in bright sunlight.	
P & IDs/ REFERENCE DOCUMENTS	D.1220.1 Section 3.2 of Volume II of Process ABC	
COMPANY/PLANT	ABC Refining and Chemical Houston Plant	
MODULE NUMBER	200.2.21.3	Page 1 of 2 March 3, 1996

Module Name

The module name should accurately and completely describe the purpose of the module. It should also be unique; if possible, no other title name should be identical, or even similar. The name should be as descriptive as possible and should contain as much detail as possible. "*Starting P-100, The Recycle Charge Pump to the Lights Column, C-121*" is more helpful than just "*Starting the Charge Pump.*"

There should be a single, consistent naming system for the modules. For example, it is inconsistent if one module is named

"Starting P-100, The Recycle Charge Pump to the Lights Column, C-121" and another, similar module is named *"How to Start P-350, V-240 Feed Pump."*

It is suggested that the module titles should have the following form:

1. The action ("Starting"). Avoid the use of words in the imperative tense on this line because instructions should not be in the Title Block.

2. What is being acted upon ("P-100, The Recycle Charge Pump").

3. What other items are affected by the action ("The Lights Column, C-121").

Note that both name and number for equipment items are provided. In the body of the procedure, it is usually satisfactory to use whichever is most commonly used by the operators. However, if there is any possibility of confusion, both should be used.

Purpose Of This Module

The next part of the title block provides a purpose for the module. Appropriate descriptive material can be provided here. Usually this section does not need many words (if it is needed at all); the module name will usually provide sufficient detail. If it is necessary to provide an extended description of the purpose of the module, reference should be made to the appropriate section of Volume II, the Operating Information volume.

Special Safety Considerations

Some operations have special safety requirements, such as the use of special clothing or the need to communicate with another department as the action is being carried out. These special requirements are in addition to normal plant safety procedures and have their own section in the Title Block. In Figure 7, no special safety issues apply. However, there is a reminder that this is a hydrogen sulfide (H_2S) service, and the standard precautions for H_2S should be in place. The same comments apply to the LPGs.

Discussion

The discussion block can be used for additional discursive material on issues such as human factors or troubleshooting.

P & IDs/Reference Documents

It is useful to show the pertinent P & ID numbers and information concerning other reference documents in the Title Block. Essentially, this serves as an index. Generally this information will be located in Volume II, the Operating Information.

Company/Plant

This section of the Title Block provides detail as to which particular process and plant-site is covered by these procedures. This is important because, once a company standardizes on a design, the procedures from different units will all look very similar to one another, and it is important to make sure that they do not get mixed up.

Module Number

The modules should have a clearly defined numbering system. In this example, the number 200.2.21.3 has the following meaning.

200	indicates area 200 of the plant
2	means that this is a start-up module
21	this is the twenty-first module in the sequence
3	shows that this is the third revision of this module

Page Number/Date of Issue

The page numbering sequence is tracked in this block. This particular module has two pages of text. Generally, it is not desirable to have an overall Volume I page numbering scheme because, if a module is added or inserted, or if an existing module adds or loses a page, the page numbering scheme will be thrown off. Indexing is carried out using the module numbers.

If the module extends to more than one page, it is not necessary to repeat all of the Title Block information on subsequent pages; it should be sufficient to reiterate only the name of the module, its number and the special safety considerations.

Conventional page numbers are not used in Volume I because the system will be interrupted if modules are added or removed. Operators will move around Volume I by using module numbers.

THE OPERATING INSTRUCTIONS

The next section of the module consists of the actual operating instructions. The instructions can be laid out using a set of columns, as shown in Figure 8. As with the Title Block, in most cases it will not be necessary to use all of these columns. The number of columns can vary within a single set of procedures depending on the content of each module.

Figure 8

Example of Operating Instructions

STEP	PER-SON	ACTION	RESPONSE DISCUSSION	REV	√
1	B-OP	Check that the level in V-100 sightglass is at least 30%.	Do not start P-100 until the 30% level is reached.		
2	B-OP	**CLOSE THE BLOCK VALVE ON THE OUTLET OF P-100**		**3**	
3	A-OP	Place FRC-121 in manual	This may cause a small temperature swing in C-121.		
4	A-OP	Manually set FRC-121 to 25% open			
5	B-OP	Open the bypass valve around FCV-121 two turns			
6	B-OP	Start P-100	The discharge pressure should be 125 psig.		
7	B-OP	**OPEN THE BLOCK VALVE ON THE OUTLET OF P-100**		**3**	
8	A-OP	Put FRC-121 on automatic			
9	B-OP	Adjust the set point of FRC-121 for a flow of approximately 2000 lb/h			

Step

The first column contains the step number. The order in which the instructions are listed is usually critical: they must be carried out in the sequence shown. If this is not the case, this should be made clear.

In order to reduce the chance of an operator inadvertently skipping a step, alternate instruction lines can be highlighted in contrasting shades or colors.

Person

The second column identifies the person who has to perform the action. Shorthand symbols can be used. In Figure 8, A-OP stands for the "A" operator, who is inside the central control room operating the Distributed Control System (DCS.) B-OP stands for the "B" operator, who is responsible for outside activities. In some cases, more than one person will be listed in this column. To save space, this column can be removed, particularly when all of the actions are to be performed by just one person. If this column is removed, the person(s) responsible for the actions can then be identified in the title block. The assignment of responsibility is particularly important when working with operators from other units or with outsiders, such as truck drivers.

Action

The action column tells the operator what to do. The following guidelines are recommended.

1. Only one action is listed for each instruction. If there are two or more action steps, they should be separated and each given its own row in the module.

2. All instructions should, where possible, be single sentences.

3. All instructions should start with a verb in the imperative tense.

4. Qualitative adverbs such as 'slowly,' 'gradually,' 'soon' and 'quickly,' should be avoided. It is much better to use quantitative terms such as 'five minutes' or 'ten degrees per hour.'

5. Avoid tangled, complex syntax or obscure words ("eschew obfuscation.")

To illustrate the above points, consider the following instruction:

> "Next, after measuring the reactor temperature, and making sure that is not above 210°C, open the discharge valve."
>
> This should read:

1. Measure the reactor temperature
2. IF it is above 210°C, go to Module XYZ
3. ELSE open the discharge valve

Response/Discussion

The Response column describes the response of the system to the action that has just been taken. For example, if the instruction calls for a pump to be started, this column can provide information as to what discharge pressure, or what discharge pressure range, is to be expected. If the observed pressure is outside this pre-defined range, the operator knows that corrective action must be taken. This is an extremely useful on-going audit step.

This column can also be used for general discussion and explanation. For example, it is a convenient place to provide discursive information that may be useful to an inexperienced operator but that the experienced operator does not need. Simple trouble-shooting information can also be placed in this column (in addition to what is provided in Volume IV, the Troubleshooting Manual.)

Revision Number

The current revision number can be placed in the revision column on those lines which have changed. This allows the operator to quickly identify the changes when a new module is issued. Another way of highlighting changes is to shade the new words (analogous to the use of cloud on P & IDs to show what sections have been modified.) Old revision numbers are discarded. In the example, Steps 2 and 7 have been modified as part of Revision # 3. They are identified by the shaded background and capitalized lettering indicate which items have been changed.

There is no reference to revision numbers 1 or 2. They have all been incorporated into the body of the procedure prior to the issue of Revision # 3. The same will happen to #3 when it is superseded by Revision #4.

If it is important to save space, this column can be eliminated. Details of the revision can then be located in the Title Block. Alternatively, certain colors or fonts can be used to indicate which words have been revised. Once the current revision is replaced by the next revision, those words that have been highlighted for this revision should be returned to their normal font and color.

Check Lists/Check Block

This column is symbolized by the check mark.

Once the procedures have been written, it is useful to develop check lists, particularly for batch operations, and for the start-up and shutdown of steady-state processes. It is not generally practical to place the check lists directly in the procedures because the manual is usually too bulky. Also it would not be possible to file or discard the check list after each use because this would destroy the physical integrity of the manual. Therefore, check lists will often be stand-alone documents that complement the operating manual.

Check lists provide a means for the operator to communicate with his supervision. For example, he might note that there was a difficulty with one of the steps in the module, although it was not of such severity to prevent the action sequence from being continued. In this case, the operator may note that the pump cavitates while he is carrying out Step 5 of the procedure in Figure 8 ("Open the bypass around FCV-121 two turns.") Therefore, he could write on to the check list that he had noticed a problem.

This is also a way in which the operator can suggest improvements to the procedures themselves.

The use of check lists to find problems with the plant or the procedures, and to communicate suggested improvements, is an excellent means of implementing both the Audit and the Employee Participation steps of the Process Safety Management standard.

It may be preferred to initial this column rather than just using a check mark.

THE AUTHORIZATION BLOCK

The third section of the module provides the authority associated with its issue. Figure 9, which is an example of this block, has five sections. As with the previous blocks, it is unlikely that all the sections will always be needed.

Figure 9

Example of an Authorization Block

REVISION NUMBER	**3**	
REVISION PURPOSE	**The block valve on the outlet of P-100 has been changed from a globe to a ball valve**	
DATE OF REVISION: July 5, 1996	WRITTEN BY:	
SIGNATURE 1 Operator		Date:
SIGNATURE 2 Operations Supervisor		Date:
SIGNATURE 3 Operations Superintendent		Date:

Revision Number

The module in this example corresponds to Revision Number 3. This does not mean that all the other modules are re-issued at the same time. It is quite possible that other modules in this set will have different revision numbers because the revision applies only to that module, not to the complete manual. The location of the sections that have changed as a consequence of the revision is shown by putting the revision number in the appropriate column of the instructions table (as shown in Figure 8.)

Revision Purpose

The purpose block describes the rationale behind the latest revision. In this case, the latest revision — Number 3 — had to do with changes to one of the valves that are used in this particular operation.

Date of Revision

The date of revision is the date of *issue*, not the date on which the module was *written*. Temporary procedures should include an expiration date. This line could also contain the date by which the module must be revised. If the company is following the OSHA Process Safety Management standard, the module will require a review every 12 months.

Written By

The technical writer/interviewer's name can be provided in the Authorization Block.

Authority/Signature Blocks

One or more of the managers must sign off on the procedure before it can be officially released. Their names and signatures go here, along with the signatures of the operator who provided the information and his supervisor.

The signing of procedures is often a major bottleneck in the whole process. There are usually two reasons for this.

1. Too many people are asked to sign the procedure. Therefore, it simply takes a long time for the procedure to wend its way

through the system. The stated reason for having many signatures is that multiple checks are put into place. However, if the issuance of the procedure is seriously delayed, safety and efficiency are compromised. As with other issues to do with procedures, a balance between thoroughness and speed must be struck.

2. The people who know the unit well, and whose signature is therefore of the highest importance, are also the people who are likely to have the most to do with other projects.

Because the signing-off process can be so time-consuming, it is important to give it considerable thought. Basically, there are three types of signature authority.

1. The first signature is from the person(s) who actually wrote the material. This could be both the subject matter expert and the technical writer (if used.) Their signature states that they have prepared a set of instructions to the best of their ability, and that the instructions are, as far as they can ascertain, accurate, complete and useable.

2. The second signature comes from someone who knows the process being described extremely well indeed. Often this will be a supervisor or senior operator. Their signature states that they have reviewed the written procedure and that, as far as they can determine, it is accurate and complete.

3. The third signature comes from a manager. He or she is not likely to know the process well enough to comment on the technical content. What their signature states is the management systems for writing procedures are in place, that these management systems address all appropriate regulations and standards, and that the systems were followed during the execution of this assignment.

Of these signatures, only the second — that of the experienced supervisor — ought to take much time.

Authorization Sheets

In order to save space, the authorization blocks can be removed from the individual modules and put on a single sheet, as shown in Figure 10.

Figure 10

Authorization Sheet

Module Number	Rev	Module Title	Plant Manager (with date)	Operations Manager (with date)	Date of Issue
1	1				
2	3				
3	1				
4	2				
5	1				
etc.					

COMPLETE MODULE

The completed module for this example is shown in Figure 11. Some detail has been removed to save space. (In this book, Figure 11 occupies more than one page. In the actual manual page sizes are larger and this will not be a problem.)

Figure 11

Complete Module

MODULE NAME	*Starting P-100, The Recycle Charge Pump to the Lights Column, C-121*	
PURPOSE OF THIS MODULE	This module describes how to transfer isobutane and other light hydrocarbons containing H_2S to the Light Ends Column, C-121	
SAFETY	**Normal precautions regarding H_2S and LPG.**	
DISCUSSION	V-100 sight glass is difficult to read in bright sunlight.	
REFERENCE DOCUMENTS	D.1220.1 Section 3.2 of Volume II	
PLANT	ABC Refining and Chemical Houston Plant	
MODULE NUMBER	200.2.21.3	Page 1 of 2 March 3, 1996

ST	PERSON	ACTION	RESPONSE/DISCUSSION	R	√
1	B-OP	Check that the level in V-100 sightglass is at least 30%.	Do not start P-100 until the 30% level is reached.		
2	B-OP	**CLOSE THE BLOCK VALVE ON THE OUTLET OF P-100**		3	
3	A-OP	Place FRC-121 in manual	This may cause a small temperature swing in C 121.		
4	A-OP	Manually set FRC-121 to 25% open			
5	B-OP	Open the bypass valve around FCV-121 two turns			
6	B-OP	Start P-100	The discharge pressure should be 125 psig.		
7	B-OP	**SLOWLY OPEN THE BLOCK VALVE ON THE OUTLET OF P-100**		3	

Figure 11 (continued)

8	A-OP	Put FRC-121 on automatic			
9	B-OP	Adjust the set point of FRC-121 for a flow of approximately 2000 lb/h			

REVISION NUMBER	**3**	
REVISION PURPOSE	**The block valve on the outlet of P-100 has been changed from a globe to a ball valve**	
DATE OF REVISION: 7/7/1996	WRITTEN BY:	
SIGNATURE 1 Operator		Date:
SIGNATURE 2 Operations Supervisor		Date:
SIGNATURE 2 Operations Superintendent		Date:

If there is no information to put into a block, then there is no need to include that block at all. For example, if there is nothing to put into the *Discussion* section of the Title Block, there is no point having that block printed at all; it may as well be dropped out for that module.

Limitations of Modularity

Although one of the major justifications for a modular approach is reduced maintenance, it should nevertheless be recognized that perfect modularity is not likely to be achieved, and that a change to one small part of the system may affect other parts in ways that are difficult to determine ahead of time. Therefore, although writing procedures in a modular fashion will reduce the effort needed to update and improve the procedures, changes should still be viewed globally as well as locally.

An example of this often occurs with process utilities. An earlier example in this chapter described a situation where a pump driver is changed from an electric motor to a steam turbine. Apparently the only modules that are changed are those that concern the normal operation, start-up and shutdown of that pump. Yet, having a new turbine on the steam header may lead to steam availability problems that have system-wide effects and that affect many other equipment items. This will mean that the modules for all the other affected equipment items may require updating.

GRAPHICS

Up to this point, the modules have consisted only of text. Where possible, they should be supplemented with graphics. Not only do graphics provide additional information and explanation, they also make the manual more attractive. This is important because a manual that is attractive and easy to read is more likely to be used. The following types of graphics can be used in Volume I.

- Photographs
- P & IDs
- Personalized Sketches
- Iconic Flow Diagrams
- Maps/Plot Plans
- Calculation Sheets (usually to go with batch operating procedures and/or checklists)
- Equipment Drawings
- Column Temperature and Pressure Profiles

PHOTOGRAPHS

Photographs are most effective when describing equipment detail. For this reason, they are particularly useful for operations involving maintenance. A photograph is completely realistic and requires a minimum of interpretation. This is particularly useful if the procedures are being used by someone who is new to the process or equipment.

Photographs cannot be photocopied directly because the images on them are of varying intensity, whereas a photocopy consists of black and white dots — it can only produce the effect of gray by changing the ratio of the dots. If it is necessary to make a photocopy of a photograph, it must be half-toned. This is done by using a half-tone screen, which converts a photograph into a fine grid of black and white dots.

In order to put text and arrows over a photograph, it is suggested that they be printed on to a clear laminated sheet which is laid over the picture. This is then copied, and the overlay becomes part of the final picture. If the picture has been scanned into the document, arrows can be added using graphics software.

Photographs sometimes contain detail that can make the overall impression confusing. It also makes management of change more difficult because the background detail may be overlooked. For example, the photograph may be of Pump, P-100. But, in the background, is a Vessel, V-100. If V-100 changes, it may not occur to anyone to change out the picture of P-100 in its module.

Figures 12 and 13 provides two examples of the use of photographs. Figure 12 illustrates operating instructions. Figure 13 is informational: it shows the location and appearance of two sets of heat exchangers.

Figure 12
Use of Photographs to Illustrate
Operating Instructions

**Do Not Allow
Level
To Fall Below
Bottom Of
Second Glass**

Figure 13
Use of Photographs to Illustrate
Operating Information

E-112, Hot Effluent/Cold Feed Exchanger

E-110, Light Ends Stripper Reboiler

P & IDs

A P & ID (Piping and Instrument Diagram) provides detailed engineering information such as line sizes, material specifications, insulation requirements and all control loops [7]. However, they are not to scale and they do not show layout. They can be used in operating procedures to illustrate complex operations, such as the line-up of a certain flow path where there is a complex network of piping. They can also be used where detailed information is needed, such as in the preparation of blind lists.

Generally, however, P & IDs are too detailed for day-to-day operations. They are best included in the Process Information Manual, Volume II.

Iconic Flow Diagrams

Most people who work on plants often prepare a personalized set of sketches and drawings that provide information on exactly what they need to do their job. These sketches are usually an informal mix of Block Diagrams, Process Flow Diagrams, P & IDs and the Process Plot Plan.

Sketches such as these are completely unofficial and unauthorized; therefore they will not have been reviewed for accuracy or completeness. Indeed, it is almost certain that many items will be missing from them because the person who drew it will have written down only those items that were of particular interest to him. For example, the sketch may shown just one valve in a certain line because that was the only one of interest to that person. If there are, in fact, two valves in that line, then an operating instruction that used this sketch concerning the isolation of the line could be mis-interpreted, and the operator might close the wrong valve.

The writer of an operating manual is faced with a dilemma regarding personalized sketches. On the one hand, there is no question that they are very useful in operating the plant, and so should be included. Yet, their lack of engineering integrity could lead to errors and confusion if they are incorporated into the operating manual. This means that the review of these sketches before they are included in the manual has to be very thorough.

One solution to this dilemma is to develop *Iconic Flow Diagrams.* An Iconic Flow Diagram is basically a P & ID with some process information added, and with detail regarding many of the minor items omitted. It helps an operator to follow the principal flows through a unit or to follow the path of a start-up. It has more engineering detail than a Personalized Sketch but less than a full P & ID. Most important, it is validated both by the engineering and operations personnel.

Iconic flow diagrams are often very skimpy. In particular many, indeed most, valves, lines and instruments will be missing. On the other hand, some items may be shown in great detail because they were of particular importance to the person preparing the diagram.

Diagrams such as Figure 14 can be invaluable in situations such as this. Valves that are to be closed are shown in solid, valves that are to be open are shown in clear. Alternatively, color can also be used to identify which valves on a drawing should be opened or closed.

Figure 14

Example of an Iconic Flow Diagram

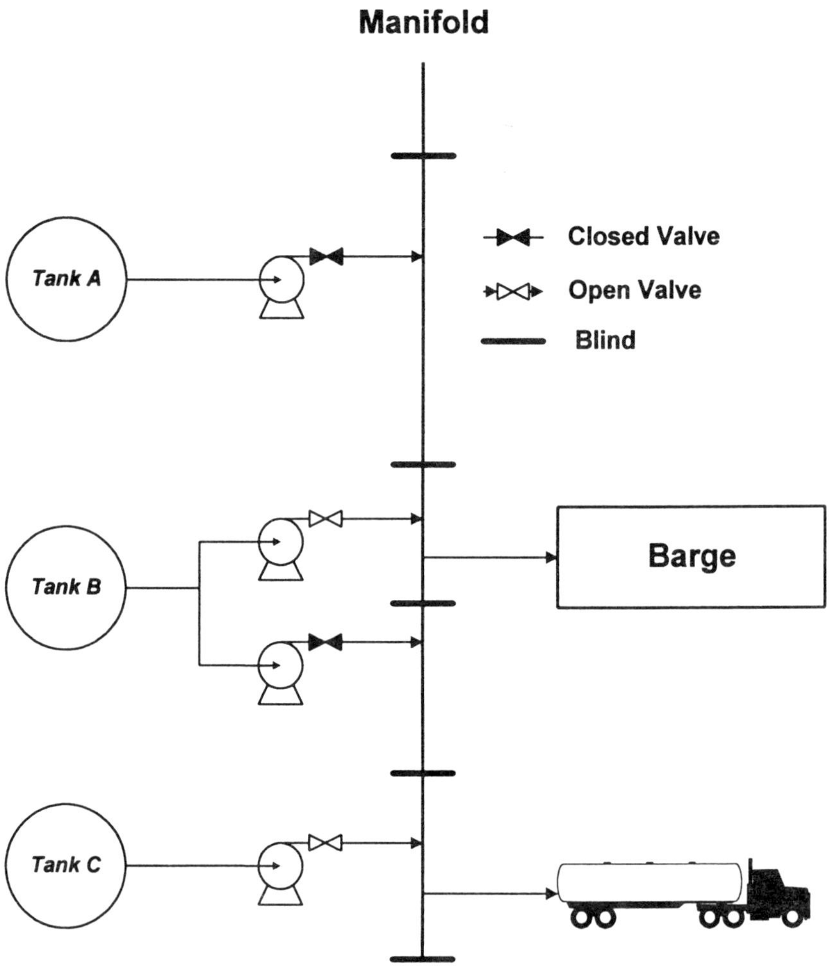

MAPS/PLOT PLANS

Maps and plot plans are particularly useful when describing any operation that requires complex valving decisions, such as unloading a truck of chemicals into a large tank farm that handles a wide variety of products through a limited number of manifolds and headers. Figure 15 is an example of a process plant map.

Figure 15

Example of a Process Plant Map

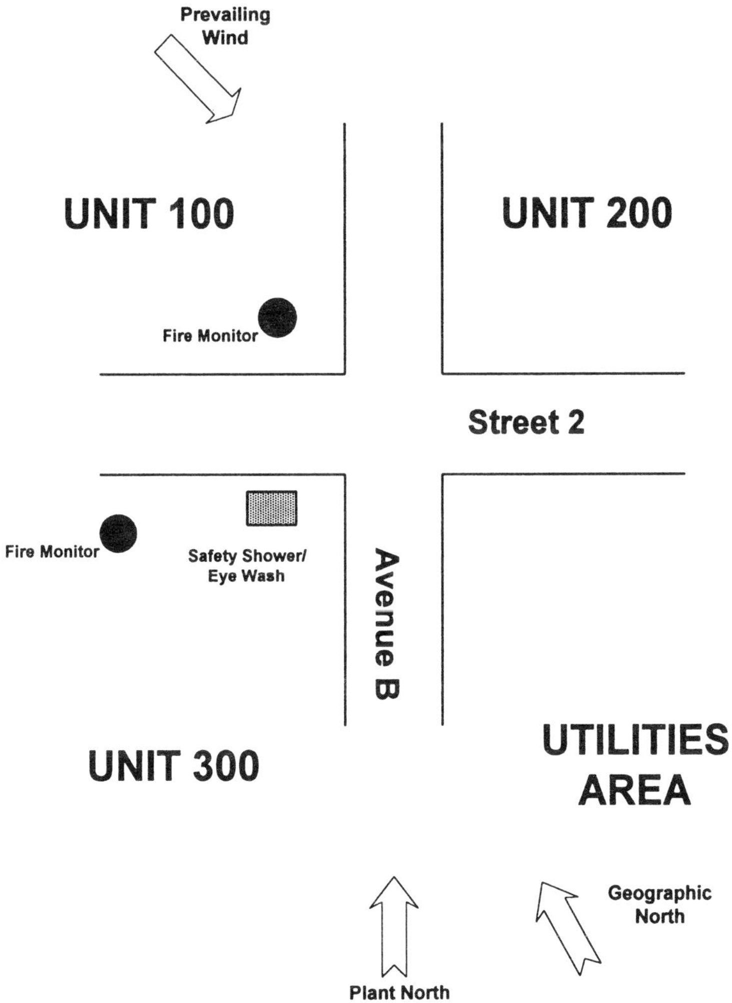

Maps are used when it is important to locate a piece of equipment. They are particularly useful for showing the location of safety equipment, such as fire hoses and first-aid stations. Maps are also handy for operations involving vehicle movement. If, for example, a truck of hazardous chemicals arrives at the gate of the facility, the security guard can provide the driver with a map of the roads to the location where he is to unload. Maps of piping systems, particularly those that are underground or that are in overhead racks, are also valuable, particularly when trying to locate a leak. Maps are useful when showing the layout of safety equipment such as fire monitors and extinguishers.

As with other examples of iconic flow diagrams and maps, Figures 14 and 15 are not intended to be complete. For example, the valves at the barge and the truck in Figure 15 are not shown.

Layout — Two Page Modules

Facing Pages

When an operator opens his manual, he is looking at two pages. In other words, as pointed out by Weiss [8], all manuals are written in modular form, whether the writer realizes it or not, with each module consisting of two facing pages. Therefore, it makes sense to organize the operating instructions so that the instructions associated with one module fit on to the two facing pages. If there is too much material, then it can be broken down into two smaller modules.

In practice, it is not always appropriate to follow this rule strictly; it sometimes makes sense to have modules use up a few pages. Nevertheless, the idea of making the information fit the layout rather than the other way around is well worth considering. In practice, if there are more than two pages of text in a single module, then the module is probably too big and should be sub-divided.

One way of achieving the goal of two-page modularity is to have a page of text and a page of graphics facing one another. The text is put on the right-hand side because it is where most readers automatically look for written material (it is how single-page reports are laid out, for example.) Figure 16 illustrates this concept.

Figure 16

Example of a Two-Page Module

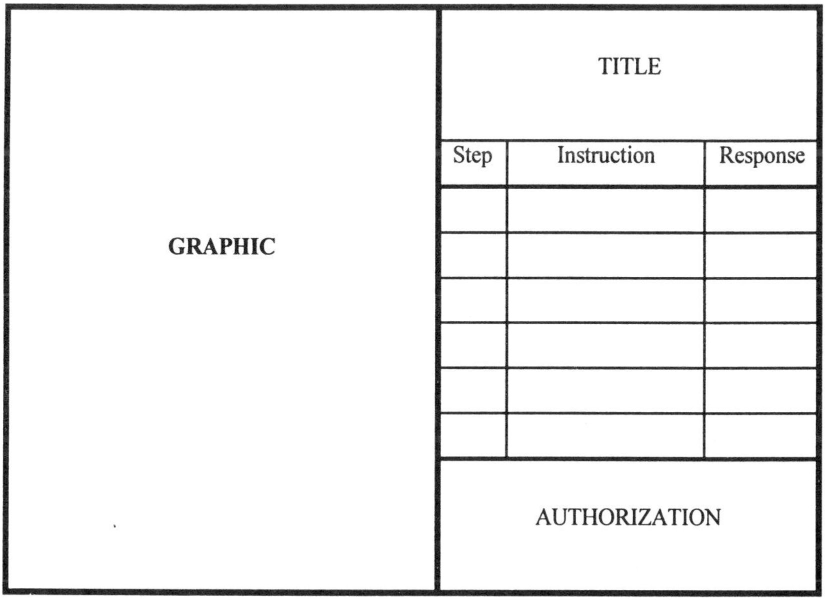

Four-Page Setup

Physically, if the two pages shown in Figure 16 are facing one another, that means that there will be two outer pages. (A *page* refers to one side of paper, a *sheet* is the physical piece of paper that has two pages — front and back.) Therefore, each module actually consists of a total of four pages — with the two interior pages containing the instructions and graphics to do with that module.

- Page 1 is the title. It includes the module number. (Putting the title here can avoid the need to also put a title block on the text page, thus saving space.)
- Page 2 is the graphic.

- Page 3 is the text.
- Page 4 is left blank (it can also be used for over-flow text.)

This approach means that the module does not share paper with other modules. Therefore, if the module changes, no other module has to be physically reproduced.

INTEGRATION OF TEXT AND GRAPHICS

One of the reasons for the text-graphic layout shown above is that traditionally it has been difficult to integrate the two on the same page. The page has to be formatted with space for the graphic, the graphic has to be inserted in that space, and then the combined elements have to be copied. This is a labor-intensive process.

Modern desk-top publishing software, and some of the more advanced word processors, now make this considerably easier. It is now relatively straightforward to drag an image into the text, and then to lay it out electronically. It still takes more effort than simply using two pages, but it is much easier than the hand lay-up approach. This being the case, it may be decided to integrate the graphics with the text. If this is done, the left-hand page will either be blank or else the procedures will be copied two-sided, with text and graphics on both sides of one-sheet, two-page modules.

ALTERNATIVE LAYOUTS

The layout shown in Figure 16 is only one way of presenting procedures. Figures 17 shows an alternative. It still uses a structured, tabular approach. However, there is no limit on the number of pages of text that can be included in one set of instructions.

Figure 17

Alternative Format

SHUTDOWN OF THE FINAL PRODUCT COOLER

Procedure Number: 5.3.12 *Page 1 of 3*

Date Written: 3/ 3/1994 Written By: A.B. Smith
Approval Date: 3/10/1994 Approved By: B.R. Robinson

SAFETY AND HEALTH CONSIDERATIONS

Chemicals used in the process: Concentrated (98%) Sulfuric Acid
Methanol
Propylene

Read MSDS to obtain:

- Properties and hazards of chemicals used
- Precautions necessary to prevent exposure
- Personal protective equipment
- Control measures if physical contact or airborne release occurs

Location of Procedures and MSDS: Area Foreman's office

CAUTION

Class "B" protective clothing must be worn during this operation.
There must be two people present at all times:
One person following these instructions
One person on fire and safety watch

Step	Instruction	Initial	Time
1	**Stop flow of acid to the recycle exchanger, E-3102**		
2	**etc.**		

Layout — Continuous Text

Although a two-page modular approach has the advantages described, some companies may nevertheless choose to adopt a more traditional discursive text style. Basically, the sequence of events for a major operation, such as the start-up of a Crude Unit, is written down sequentially. A single instruction set typically is about 5-10 pages in length, although it is possible to have operating procedures of over 100 pages.

In spite of the differences with the modular approach, many of the techniques are the same. For example, each page will have a header describing what instructions are being given. Each page will also have a footer providing authorization and dates.

The layout shown in Table 3 is widely used. It is based on the style often used in nuclear power plants. This type of approach is very structured and complete. It will satisfy audits well. However, it does tend to be voluminous, and does not satisfy the principles of minimalist writing (see Chapter 6.)

Table 3

Continuous Text Layout

1.0 PURPOSE

Describes the purpose of the procedure, and the scope of operations.

2.0 SPECIAL CONSIDERATIONS AND PRECAUTIONS

Describes any special safety issues and/or instructions.

3.0 PREREQUISITES

Describes any actions that must be carried out before the operations in this section can start.

4.0 REQUIRED EQUIPMENT

This is self-explanatory

5.0 REFERENCES

This section mentions any documents that should be referred to, particularly as they affect safety.

6.0 PROCEDURE

The actual procedures are described here. Generally, there are multiple levels: 6.0, 6.1, 6.1.1, etc.

START-UP PROCEDURES

In the Introduction to this chapter, it was suggested that Volume I could contain five sections, of which the first was Start-Up.

THE STARTING POINT

Before writing start-up procedures, an assumption has to be made as to what condition the plant is in at the point of start-up. There are various possibilities.

1. Major turnaround activity has just been completed on the plant. Equipment is still open for entry and blinds are still in place. Electrical equipment is still locked out. Instrument loops may or may not be in service.

2. Maintenance has been completed on the plant; the plant has been buttoned up, with all blinds and lockouts removed. However, the equipment contains air and possibly small amounts of water from the clean-out following the turnaround work.

3. The plant is has been inventoried with hydrocarbons and all equipment is ready to be operated. This is the normal condition before a start-up if the plant was shut down for reasons other than major turnaround.

4. The plant is on stand-by. All equipment is operable and inventoried with the correct process materials in each piece of equipment. This is the condition following a temporary shutdown lasting for just a short period of time. Start-up from this condition is sometimes referred to as a "hot start-up." Another term sometimes used for this condition is "static operation."

It is the second of these four conditions that are used in this book as the starting point for a start-up, i.e. control of the plant has passed from maintenance to operations, but the plant has yet to inventoried with process chemicals, and all equipment items have to be started from a cold condition. In particular, the plant has not been purged of air and water. Start-up from Step 1 is described in Chapter 5.

START-UP TIMELINE

A start-up is a project, often lasting for a few days. It is important that it be managed well, otherwise there could be substantial losses due to lost production and damaged equipment. A badly organized start-up is also likely to be more hazardous to the people working on it. It therefore makes sense to develop a critical path chart, or other project management schedule, with which to organize the start-up. This should be included in Volume I.

SHUTDOWN PROCEDURES

Procedures for (planned) shutdowns have the same features as start-up instructions, except that the starting point is usually assumed to be that the plant is in full operation, and the end of the sequence of operations results in having the plant ready for maintenance. However, there may be significant variations on this. In particular, equipment availability may be restricted. For example, the reason that the plant is being shut down may be that an item of equipment, such as a pump seal, has failed. Therefore, the shutdown instructions may have to be modified so as to reflect the absence of this pump.

LEVELS OF SHUTDOWN

Just as there are various starting points for a start-up, so a plant can be shut down with varying degrees of completeness. For many plants, there are three levels of shutdown.

Stand-by

When the plant is on stand-by the unit continues to operate. However, there is no feed coming in and no product being made. As far as possible, conditions are kept as close to normal as possible so that the plant can be restarted with minimal trouble. Distillation columns are put on total reflux, vessels are operated at their normal level and rotating equipment is kept running. Only the items that are actually being repaired are taken out of service.

Stand-by is used when there is a short repair job to be carried out that does not require a large maintenance effort.

Normal Shutdown

A normal shutdown is that state where all rotating equipment, heaters and other unit operations are stopped. No attempt is made to keep the unit in a semi-operating state. However, equipment is not de-inventoried, nor is anything purged — except for the equipment that is being worked on.

Turnaround

When a plant is shut down for turnaround, it is completely deinventoried and purged. Blinds are installed and instruments may be disconnected. Generally all vessels are made ready for maintenance and personnel entry.

SHUTDOWN HAZARDS

One important difference between shutdown and start-up is that, if something goes wrong during a start-up, particularly in the early stages, it is possible to stop further action and to take time to correct what has gone wrong. This is not true for shutdown. For example, if the first step in the start-up of a distillation column is to put feed into the tower, a failure of the feed pump will mean that the tower will remain in a de-inventoried condition. With shutdowns, the opposite applies. Once the sequence of shutdown events is initiated, it can be difficult to stop and there may not be much time to correct problems. Consequently, potentially dangerous situations, such as stopping feed to a fired heater before shutting it down, may occur.

This is analogous to flying an airplane. If, during the "start-up" of an airplane something goes wrong, such as the engines fail to start or the brakes will not release, the plane will remain on the ground in a safe condition until the problem is fixed. On the other hand, if the plane is coming in to land and the engines fail, or if the brakes fail after landing, an accident will ensue.

Normal Operations

Routine Rounds

For a continuously operating plant, normal operating procedures are often quite simple. There is a routine of taking readings, catching samples, filling out log sheets and carrying out special instructions that does not vary much from day to day: a routine that is not difficult to carry out if everything is running smoothly. Therefore, the written procedures for normal operations do not generally require linkage. In effect, the normal operating procedures are like an extended check list.

These procedures also include descriptions of shift handover, logbooks and logsheets.

In the case of batch operations, the normal operating procedures are in effect start-up and shutdown instructions that are repeated many times, often with a checklist included.

Routine Tasks

Another part of Normal Operations includes the routine tasks that an operator is expected to carry out on a non-scheduled basis. Some of these tasks, such as how to catch a sample or how to switch pumps, are equivalent to the Generic Procedures. Table 4 provides a list of typical routine tasks, many of which are carried out on a daily or weekly basis.

Table 4

Example of Routine Task List

Action	When	Discussion
Safety		
Relief Valves	Sunday	
Rupture Disks	Sunday	
Fire Monitors	Sunday	
Fire Extinguishers	Sunday	
Fire Hoses	Sunday	
Safety Showers	Sunday	
Eyewash Stations	Sunday	
Air Packs	Sunday	
Face Masks	Sunday	
Bunker Gear	Sunday	
Work Orders/Permits		
Safety Permits	Daily	
Maintenance Work	Daily	
Environmental		
Sewers	Every four hours	
Storm Water Drains	Every four hours	
Flare	Every four hours	
Equipment		
Lube Oil Pressure/Cleanliness	Every four hours	
Seal Gas	Every four hours	

TEMPORARY OPERATIONS

Procedures for temporary operations are very important because these operations are often high risk. In spite of their importance, it is frequently an area of manual-writing that gets overlooked. By their very nature, temporary operations are difficult to write procedures for because they will not have been identified prior to their being needed. Temporary operations can often be identified by their frequent use of words and phrases such as “hot-tap,” “flex-hose,” “overtime” and “vacuum truck.” These operations are usually of short duration.

A temporary operating procedure has to be written if, for example, the body of a control valve starts to leak. Management may decide to isolate the valve and run on the control valve’s bypass only. This is a temporary operation, and may last for only a short period of time. Nevertheless, a procedure for this operation must be prepared and included in Volume I of the operating manual. Because the contents of this section of the manual are temporary, it will normally contain very little material. However, a portfolio of temporary operations will build up over time. If the temporary operation is likely to recur, management may decide to set up a file of inactive procedures for temporary operation that can be reactivated when necessary.

The temporary nature of this type of work can generate a “Let’s just get on with it attitude.” Statements such as “It will take longer to write the procedure than to do the job” are often made, and such statements may very well be correct. Nevertheless, it is critical that this attitude does not prevail, and that the procedures do get written before the work is allowed to start.

Once the temporary operation has been successfully completed, its modules must be removed from the manual, unless it is intended to make the operation permanent. Removing the procedure shows that there is no longer any authority to operate the plant using the temporary operation. Temporary operating procedures should always include their own expiration date; indeed this is one way in which they can be identified.

Some operations, such as the unloading of a rarely-used chemical, fall into the category of *permanent temporary.* That is, the operation itself is temporary because, for example, special piping may have to be installed. However, the operation is certain to be repeated at some time in the future, therefore it is permanent. For this reason, a permanent operating procedure for it should be developed and incorporated into the manual.

A related category is *rarely-used permanent.* In this case, the operation is permanent in the sense that it does not require any type of process or equipment changes. However, it is rarely used. In both of these cases, permanent operating procedures are required.

One other set of operating modules that can be conveniently included in the temporary operations are those that describe the preparation of equipment before handover to maintenance, and the re-commissioning of the same equipment when maintenance has finished their work. Examples include the draining and purging of pump cases, and the removal of relief valves for inspection and bench-testing. While such operations are not strictly speaking "temporary" because they are, after all a part of the permanent fabric of the plant, they are not really operating procedures either, so it is reasonable to include them in this section.

Generic Procedures and Vendor Manuals

Generic Procedures

Many procedures are generic in nature. For example, the procedures for starting pumps, lighting furnaces or catching liquid hydrocarbon samples are likely to be very similar at different points in the plant. Therefore, it makes sense to develop a set of generic procedures for these operations, and to assume that the operator has been trained in their contents. Then, the instructions for the particular items do not need to repeat the common material; instead they can concentrate on those aspects of that particular operations that are different from the generic operation.

When there are duplicates of equipment items, it is tempting to write one set of procedures for both. While there is nothing wrong with this in principle, it is very important to make sure that the two items really are duplicates of one another, and that there are no detail differences between them that materially affect the operating procedures.

Vendor Manuals

A common source of information, particularly for the upkeep and maintenance of equipment, is vendor manuals. These are provided by the vendor when the item in question is purchased. Such manuals are usually completely generic in nature. They are likely to be of a completely different format from the company's own procedures, and they may cover more components than are actually present in a particular case.

Level of Detail

Deciding on the level of detail that is needed in the procedures is difficult. If there is too much detail, the manual will become bulky and wearisome, and it will be ignored, particularly by the more experienced operators. On the other hand, too little detail will provide the inexperienced operators with insufficient guidance.

The use of the "ACTION/RESPONSE" columns, as shown in Figure 8, can help overcome this dilemma. The "ACTION" column contains a description of what has to be done. It can be written in a bulleted, near-cryptic manner, and thus be useful for the experienced operator. Additional discursive detail can be provided in the "RESPONSE" column.

When planning a procedures-writing project, it is very important to think through the issue of detail because it will dramatically affect the amount of time and money needed.

In practice, the problem of how much detail is needed often tends to resolve itself. During the writing process, the operators will describe the operations at the level of the detail that they need. If they are writing about the unloading of a truck of highly hazardous chemicals, the operation will be described minutely. On the other hand, procedures for checking pressure gauges in a utility system will probably be quite brief.

Manuals are generally written at one of the following three levels of detail.

Level 1 — Overview

At this level, the manual contains just broad, general statements, such as "Commission the Condenser" or "Load the Truck." Little or no plant-specific detail is provided. This roughly corresponds to the fourth level in the example of Top-Down development illustrated in Figure 5.

Level 2 — Equipment Description

The next level, which is probably the most widely used, is to provide equipment description. So the previous instruction: "Commission the Condenser" would be divided into details concerning starting the cooling water and measuring the temperature of the process streams. The module shown in Figure 11 is at this level.

Level 3 — Valve Detail

The final level, which is uncommon on process plants (although it is standard in nuclear plants) provides virtually action-by-action instructions concerning opening and closing valves, pressing DCS panel buttons and the use of instruments.

If it is decided to write procedures at this level, a valve/pump checklist such as that shown in Table 5 can be very useful. It shows the status of each control valve and pump on the unit at the conclusion of each operation for each P & ID. Even if this type of list is not included in the final product, it is useful for the procedures-writers to develop one as a check on their own work.

Although it looks complicated, its use is quite straightforward. For every step, the appropriate changes are made. A table like this is handy for catching oversights, such as forgetting to turn on a pump before filling a vessel.

Table 5

Valve/Pump Checklist

P&ID	Step	1	2	3	4	5	6	7	8
1000	P-100	Off	Off	Off	Off	Off	Off	Off	Off
	HV-1001	O	C	C	C	C	C	C	C
	LV 1001	O	C	C	C	C	C	C	C
1001	FV 1004	O	C	C	C	O	O	O	O
	LV 1003	O	C	C	C	O	O	O	O
	PV 1006	O	C	C	C	C	C	C	C
1002	FV 1005	O	O	O	O	O	O	O	O
	FV 1007	O	O	O	O	O	O	O	O
1003	FV 1009	O	C	C	C	C	C	C	C
	FV 1010	O	C	O	O	O	O	O	O
	TV 1007	O	C	O	O	O	O	O	O
	TV 1008	O	C	O	O	O	O	O	O
1004	P-101	Off	Off	Off	Off	Off	Off	Off	Off
	FV 1014	C	C	C	C	C	C	C	C
1005	HV 1008	O	C	C	C	C	C	C	C
	PDV 1038	O	C	C	C	C	C	C	C
	PV 1036	O	C	C	C	C	C	C	C
	FV 1015	O	C	C	C	C	C	C	C
1006	FV 1021	O	C	C	O	O	O	O	O
	FV 1019	O	C	C	O/C	O/C	O/C	O/C	O/C
	PV 1046	O	C	C	C	C	C	C	C
	FV 1018	O	C	C	C	C	C	C	C
1007	P-102	Off	Off	Off	On	On	On	On	On
	HV 1017	O	C	C	O/C	O/C	O/C	C	O/C
	HV 1015	O	C	C	C	C	C	C	O/C
1008	FV 1023	C	C	C	C	O	C	C	C
1009	P-103	Off	Off	Off	Off	On	On	On	On
	P-104	Off	Off	Off	Off	Off	Off	Off	On/Off
	FV 1028	O	C	C	C	O	O	O	O
	HV 1016	O	C	C	C	C	C	C	C
	PV 1057	O	C	C	C	O	O	O	O
	PDV 1058	O	C	C	C	O	O	O	O
	FV 1022	O	C	C	C	O	O	O	O

Sources of Information For Volume I

For plants that are already operating, information for the procedures can come from various sources, including:

- Experienced operators and supervisors
- Existing procedures
- Logsheets and other paper and electronic records

Interviewing The Operator

For plants that are already operating, most of the information that will be written into the procedures will come from the operators who are working on the plant. The "interview" process is described in Chapter 6.

Existing Procedures

For plants that are already operating, there is usually an existing set of procedures. These procedures may be out-of-date, incomplete and difficult to use. Nevertheless, this does not mean that they have no value; they will almost certainly contain some useful information. However, all the instructions in them will have to be checked before they are written up in the new manual. Existing procedures are often a particularly good source of background and reference material, most of which will be included in Volume II.

Logsheets, Laboratory Data and DCS Records

Logsheets are an important part of most operations, particularly for batch processes. Logsheets are generally used to record events in a chronological order. Usually, an operator fills in a standard form and enters on to it mostly instrument readings. The information is usually written in by hand. Logsheets should be used for the two following reasons.

2. The log sheet provides a record of problems as they develop. This is invaluable should there be an accident or major upset, and it becomes necessary to trace what happened. Of course, this requires that the log sheets be properly indexed and stored.

Plants which use Distributed Control Systems (DCS) can store information from variables and this information can be recorded at a much greater frequency than on a manual system. Nevertheless, even for these plants, the manual recording of data can be useful.

Log Books

Operating log books contain written information that is entered on a chronological basis in an unstructured manner. They differ from logsheets in the following ways.

1. Logbooks are used to record actions that are either non-standard or that are performed relatively rarely.

2. The information is entered in a free form manner.

Logbooks should preferably be bound, with the pages numbered in sequence. If a loose-leaf system is used, it is important to make sure that the pages cannot be mixed up, particularly when the book is being filed. The information in the book should be clearly identified as to what unit is being described, and who is entering the information. If ever there is an accident on the plant, it is important to secure the logbook immediately before it "disappears."

Batch Operations

It is extremely important to write good procedures for the operations where time is a variable, because the probability of an accident is much higher than during steady-state operations. One study [9] concluded that some two-thirds of process control problems occur during start-ups, shutdowns and batch operations. Most people with plant experience would confirm this conclusion from their own practical knowledge.

with plant experience would confirm this conclusion from their own practical knowledge.

Operating procedures for batch plants are often more complicated than for steady-state plants for three reasons:

1. Time is a variable; hence a sequence of time-dependent instructions is required.

2. Each batch may require a variation on the standard "recipe." Therefore some kind of calculation or batch sheet is usually necessary.

3. Many batch plants are designed to make multiple products from the same items of equipment at different times.

In spite of the above comments, the differences between procedures for batch and steady-state plants are not as different as might be expected because most procedures for a steady-state plant cover start-up and shutdown, which are essentially batch operations.

Many batch plants keep records of each individual batch operation as part of their overall quality management program. In these cases, it is suggested that a set of generic procedures be written and that batch or calculation sheets that correspond to the generic procedures be prepared. For example, a step in the procedure may call for the addition of an unspecified quantity of catalyst. The batch sheet is used to calculate how much is required for this particular batch. It also shows how much was actually added. Many batch sheets will include check boxes. These can call for information such as the time an action was performed. Batch sheets may also include space for calculations.

CHAPTER 3
VOLUME II — OPERATING INFORMATION

INTRODUCTION

In Chapter 1, it was suggested that the operating manual be divided into five volumes, of which the second, Volume II, is entitled "Operating Information." Volume II contains background information and reference material that the operators may require during the course of their work. By separating information from procedures into two separate volumes, both will be easier to use.

In addition to being used by the operators, much of the information in Volume II will be used to support other management and technical functions, particularly in the area of Process Safety Management (PSM). Volume II will assist with mechanical integrity programs, hazards analysis reviews and the management of change system. In effect, Volume II will satisfy many of the Process Safety Information requirements of the PSM standard.

One exception to the idea of placing all background information in Volume II may concern emergency information. During an emergency, the operators and other personnel are going to need all pertinent information in one place. Therefore, both the Emergency Operating and Shutdown Procedures and the Emergency Response Manual (both of which are described in Chapter 4) should include their pertinent reference material.

A representative Table of Contents for Volume II is provided in Table 6.

Table 6

Representative Table of Contents for Volume II

I. **OPERATOR DUTIES AND RESPONSIBILITIES**

A. OUTSIDE OPERATOR

1. Shift Change
2. Visual Checks
3. Coordination with DCS Operator

B. INSIDE OPERATOR
C. UTILITY OPERATOR
D. TASKS LISTS

II. **PROCESS DESCRIPTION**

A. PROCESS OVERVIEW

1. Unit Overview
2. Unit 100
3. Unit 200
4. Unit 300
5. Unit 400
6. Unit 500
7. Utilities

B. PROCESS BLOCK DIAGRAM
C. PROCESS CONTROL STRATEGY
D. PROCESS FLOW DIAGRAM (PFD)
E. PIPING AND INSTRUMENT DIAGRAM (P & ID)
F. ISOMETRICS/MODELS
G. GEOGRAPHICAL LOCATION

Table 6 (continued)

III. **OTHER PROCESS INFORMATION**

A. PROCESS CHEMISTRY
B. STREAM ANALYSIS
C. QUALITY MANAGEMENT
   1. Raw Materials
   2. Product
   3. Statistical Process Control (SPC)

IV. **SAFETY, HEALTH AND ENVIRONMENTAL**

A. SPECIAL OR UNIQUE HAZARDS
B. INADVERTENT MIXING
C. SAFETY SYSTEMS AND THEIR FUNCTIONS

   1. Emergency Shutdown System
   2. Deluge System
   3. Bleed and Vent Points
   4. Blind List
   5. Car Seal List
   6. Cable Locks
   7. Emergency Isolation Valves

D. ALARMS/INTERLOCKS/TRIPS
E. MATERIAL SAFETY DATA SHEETS (MSDS)
F. HAZARDOUS CHEMICALS

   1. Properties and Hazards
   2. Exposure Mechanisms
   3. Prevention of Exposure
   4. Response if Exposed to the Chemical

Table 6 (continued)

G. SLIPPING, FALLING, TRIPPING, CRUSHING

1. Types of Hazard and Their Potential
2. Prevention
3. Response

H. MATERIAL SAFETY DATA SHEETS (MSDS)

I. SAFETY EQUIPMENT AND CLOTHING

J. HOUSEKEEPING

1. General Rules
2. Inspection Teams
3. Disposal of Trash and Refuse
4. Special Disposal Rules

K. PERMIT TO WORK PROCEDURES

1. LockOut/TagOut
2. Confined Space Entry
3. Opening Equipment or Piping
4. Facility Entrance Control
5. Stewardship
6. Checking
7. Two-Way Radios

Table 6 (continued)

V. **OPERATING LIMITS**

A. DEFINITION
B. APPLICATION
C. PROCEDURES FOR CHANGING OPERATING LIMITS
D. VARIABLE VALUES AND ALARM SETTINGS

1. Unit 100
2. Unit 200
3. Unit 300
4. Unit 400
5. Unit 500
6. Utilities

E. RELIEF VALVE SETTINGS

VI. **EQUIPMENT AND VALVE INFORMATION**

A. P & IDs
B. Major Equipment Items
C. Instrument List

VII. **CONTACTS**

A. Fire and Ambulance
B. Environmental
C. Management
D. Vendors

OPERATOR DUTIES AND RESPONSIBILITIES

Volume II should describe each operator's responsibilities. Included in this information are:

I. A formal job description
II. Responsibilities, including
    A. Who to report to
    B. Area covered
    C. Functions covered
III. A check list of tasks to be executed at the start and conclusion of each shift (and at other times as appropriate)

Table 7 provides an example of a brief Task List.

Table 7

Abbreviated Task List

#	Area	Detail
1	**Safety**	Maintain the highest level of safety by observing all company procedures. This includes: • Ensure that all *Permit to Work* policies are followed exactly • Pause before opening or closing any valve and think through the consequences of that action • Ensure that all workers and visitors on the unit are wearing the correct safety equipment at all times • Report any potentially dangerous situations to management ***immediately.***
2	**Unit Tour**	Tour the unit at the start and end of each shift. Complete the check list at the start and end of each shift. Monitor housekeeping, equipment, safety equipment, lubrication, general operations, environmental and maintenance activities on a continuous basis
3	**Training**	Be aware of all training programs. Take the relevant courses. Identify any areas of the operation in which additional training is required.
4	**Environmental**	Know the pertinent environmental rules and regulations. Know the procedures for reporting incidents. Ensure that they are followed.
5	**Product Quality and Quantity**	Ensure that daily orders are followed. Catch samples according to schedule and adjust operation according to the results.
6	**Logged Information and Reports**	Ensure that all log sheets, log books and reports are completed on schedule and delivered as required.
7	**Equipment**	Ensure that all rotating equipment is properly lubricated according to the schedule.

Part of the Task List is the Unit Tour. Figure 18 provides an example of a Unit Tour structured as an operating module, as described in Chapter 2. This tour is to be carried out by the outside operator at the start and end of each shift. Lists of this type are particularly important because many accidents occur within 30 minutes of shift change (both before and after.)

Figure 18

Example of a Unit Tour Module

Module Name	***Outside Operator - Routine Rounds***			
Purpose	The following is a general listing of the Outside Operator's duties and duties. This list is not comprehensive. Rounds and duties have to be adjusted according to workload, work permits, and the weather. The items in this round must be performed at the start and the end of the shift. They should be repeated as necessary during the shift.			
Module Number	200.1.1	**Person**	Operator	**Page** 1 of 7

Step	**UPDATE SAFE WORK PERMITS AT THE START OF EACH SHIFT**
1-1	At the start of the shift, obtain a thorough verbal turnover from the previous operators concerning unit conditions and any special instructions that were given. This turnover should include a description of equipment that Maintenance is working on or scheduled to work on.
1-2	Read the turnover book back to the last shift that was worked by the oncoming operator.
1-3	Check the shift foreman's write-ups.
1-4	Go to the change room and put on the proper safety clothing.
etc.	

PROCESS DESCRIPTION

The Process Description section of Volume II will usually include the following sections.

PROCESS OVERVIEW

The process overview provides a written description of the process and how it works. The overview includes a summary of the major product flows, the feedstocks that are used, a brief explanation of the process economics and issues such as seasonal changes in product mixes, or limitations in the availability of utilities. It assumes that the reader is familiar with chemical process technology in general, but not with the particular unit that is covered by this set of operating procedures.

The process overview is often written by the technical staff; consequently there is a danger that they could provide too much material, or that the information will be too specialized. It is important to remember that procedures are intended to serve the operators' needs; the procedures are not a surrogate engineering manual.

PROCESS BLOCK DIAGRAM

A process block diagram is useful for orienting people unfamiliar with the plant. It is usually used to supplement the process overview. Each reaction and purification step should be shown, along with the streams that connect those steps. Only the more important process streams are shown; often the principal feed and product streams are highlighted. Each block in the diagram has a brief written description associated with it that explains what its primary function is. Process Block Diagrams are useful for initial training and for developing a conceptual understanding of the process, but otherwise they do not have much place in operating procedures because they are not detailed enough. They are similar to Iconic Flow Diagrams, discussed in the previous chapter.

Process Control Strategy

This section provides an overview of the principal control loops on the plant. It does not discuss every local controller, instead it focuses on the overall control strategy.

Process Flow Diagram (PFD)

At the heart of any process is the Process Flow Diagram or PFD, sometimes referred as a Process Flowsheet or as the Heat and Material Balance. For each stream, the following information is typically provided in a PFD:

- Total flow rate
- Composition by component flow rate or weight fraction or boiling fraction, with the molecular weight of each component shown
- Temperature
- Pressure
- Density (for liquids) or Specific Volume (for gases)

Other stream information items that might be included are pH, solids fraction, enthalpy, entropy and compressibility (for gaseous streams.)

The PFD provides some detail regarding equipment items. For example:

- Power requirements for rotating equipment
- Number of trays in distillation columns
- Heat duty and LMTD values for heat exchangers

PFDs are very important when learning about a plant, particularly the process details. However they are not usually an important part of the operating procedures because they do not provide a physical representation of the plant, nor do they show every line and minor piece of equipment. The PFD is usually too abstract for operating use.

Piping and Instrument Diagram (P & ID)

Piping and Instrument Diagrams (P & IDs) provide detailed mechanical and instrumentation information. Typically, the following information is to be found on a P & ID.

- All equipment items (with names and alpha-numeric identifiers). The equipment diagrams are roughly to scale and their approximate elevation to one another is provided. Important equipment internals are shown.

- All process, utility and service lines. Line dimensions are provided, along with details concerning insulation, gasketing and tracing.

- All materials of construction are identified, and all spec breaks clearly delineated.

- Special information (such as that a certain line should not be pocketed.)

- Instrument and control loops.

- All valves, including relief valves with their set relief pressures.

P & IDs do not contain process information, such as the composition of the streams in the lines and vessels (this is provided in the PFD) — nor are they drawn to scale. P & IDs are developed before the plant is laid out; consequently, they do not usually contain information on lengths of pipe or distances between items. Because conventions concerning P & ID labeling vary from company to company, the first drawing in a set of P & IDs should explain the symbols that are being used in this project.

It is important that high quality P & IDs be available at the start of a procedures-writing project because they contain all equipment detail. So, for example, every valve can be properly identified in the operating instructions. Also, every location for inserting blinds can be shown.

Whether the P & IDs are to be used in the manual depends mostly on the level of detail in Volume I. If the instructions are at the valve-by-valve level, then P & IDs will be needed. This will also require that the valves on the P & IDs be numbered.

One important role of the procedures-writing team is to make sure that the plant is operable as built. Frequently, it will be necessary to have special start-up lines, for example to dispose of the initial product which may be heavily contaminated with catalyst fines. As the manual is being written, the team will be checking for the presence of these lines.

Although a set of P & IDs could be physically included with Volume II, a better system might be to keep a set of them on a rack in the control room. If they are put in the procedures proper, they may be damaged or lost.

Isometrics

Isometric drawings are used primarily by the engineering staff; they provide a to-scale, three-dimensional image of the plant. They are not usually used much by operations personnel because they are too detailed and too cumbersome. Volume II may provide a reference as to where they can be found.

Models

When brand new plant is to be built, plastic models are often prepared to check for overall operability and maintainability of the final unit. They are very useful when the first edition of the manual is being prepared. They help the writer located items geographically and to develop an understanding of the physical scope of the operational instructions.

Models are rarely kept up to date once the plant is commissioned, so they quickly lose their value to the people who are writing the procedures.

Increasingly, models are built on 3-D CAD systems. Operating manuals of the future will have walk-through capability based on these systems.

OTHER PROCESS INFORMATION

Process Chemistry

It is useful to put some basic process chemistry in Volume II. It gives the operator an overall understanding of how the process works. However, it is important not to put too much detail here. Realistically, the operators are not likely to ever actually use this type of information.

Stream Analysis

Most processes have a routine for catching and analyzing samples. This section of Volume II describes what this operation entails. It includes:

1. A list of the samples that are to be caught
2. Where they are to be taken from
3. How they are to be handled and/or transported
4. What the reported results signify
5. What to do if the reported results are out of the acceptable range

Sampling is analogous to taking instrument readings. A sample provides snapshot information about the process for a particular point in time. Therefore, operating limits need to be established for the sample results. The target information can be provided in modules, along with upper and lower values. If the reported sample results are outside the range of those values, then the operator knows that there is a potential problem. He also knows that he is not allowed to intentionally deviate outside this range without going through the management of change process.

Quality Management

Increasingly, operators are responsible for quality management. Indeed, the whole process safety management program, can be viewed as a subset of a total quality management program. This means, therefore, that the operating procedures need to reflect the quality control issues for which the operators have day-to-day responsibility.

Statistical process control is a part of some quality programs. Volume II is an excellent place to present the background and instructions on this topic. In particular, this section can provide instructions on how to handle deviations from the control limits.

Environmental

This section of Volume II identifies each stream that is discharged to ground water or to the air. This section can contain other environmental information, such as how to dispose of hazardous materials or a list of the reports that have to be filed in the event of a spill.

SAFETY AND HEALTH INFORMATION

Volume II should contain information on the means available to the operators on how to prevent exposure to the hazardous chemicals when they are released. The usual way of doing this is with safety clothing and breathing-air packs. However, other means, such as evacuation to a sealed and protected control room, may also be applicable.

SPECIAL OR UNIQUE HAZARDS

Some plants have special hazards that have to be guarded against. Examples of these include:

1. *Exposed high voltage equipment.* Some processes, such as those that employ electrolytic cells, use high voltage equipment that cannot be fully insulated. Clearly, the operators have to be warned

of the dangers associated with contacting the high voltage equipment and what safeguards are provided to prevent an accident from happening.

2. *Unusual reactions*. Sometimes, a process will respond to an upset condition in an unusual way. For example, some monomers will, when accidentally polymerized, form a "popcorn"-like solid that can swell and physically burst pipes and flanges.

3. *Liquid full piping*. In most plants, overhead lines from columns and condensers are designed for liquid-full service, even though they usually only contain vapor. Sometimes, however, the overhead lines are so large that there is a considerable incentive to reduce capital costs by designing them for vapor contents only. That way, the pipe walls can be thinner and the supports less robust. If the operations department can satisfy themselves that they will not accidentally fill such lines with liquid, it makes economic sense to choose this low-cost option. This does mean, however, that the procedures must carefully address the problems associated with accidentally overfilling equipment and putting liquid in these lines, thereby causing a catastrophic line failure.

Inadvertent Mixing

One problem that OSHA's Process Safety Management standard specifically addresses is the inadvertent mixing of different types of chemicals, and the consequences of the accident that could ensue, were such mixing to take place. Scenarios that could cause this are frequently identified in process hazards analyses, such as HAZOPs.

HAZOPs are especially good at picking up one particularly dangerous and potentially common type of inadvertent mixing: the unexpected reverse flow of a process fluid from the process to a utility line (generally this assumes that a check valve failed.) The consequences of this scenario can be that the hazardous chemical enters a utility header, circulates all around the plant and enters equipment where its presence will be very unexpected and may be extremely hazardous.

Safety Systems And Their Functions

All plants utilize a wide variety of safety systems, i.e. systems that are installed specifically to prevent an accident from occurring, or to reduce its consequences should it already have taken place. Some examples of such safety systems include:

Pressure Relief Devices

Relief valves, or equivalent devices, such as rupture disks, are required by code on pressure vessels. It is important that the operators know where the relief devices are located, and what items of equipment each one is protecting, particularly in situations where the designers used one relief valve to protect two or more vessels. Should this be the case, the operators must understand the importance of not isolating one vessel from the other and thus separating the second vessel from its relief protection.

Motor-Operated Valves

These are commonly located around pumps where failure of a seal could lead to a catastrophic release of toxic gases.

Alarms

Alarms are triggered when a variable goes out of the permitted range and when this deviation is considered important enough to justify the cost of the instrumentation and the immediate attention of the operators. Usually, an alarm generates a visual signal, such as a flashing light, and an audible signal, such as a horn. The operator acknowledges the alarm by pressing a button. This stops the light flashing and it silences the horn; it does not, however, turn off the light itself. The problem has to be corrected before this will happen.

Interlocks and Trips

Most plants make extensive use of interlock and trip systems to protect against hazardous conditions. Although the word "interlock" is often used for both terms, there is a difference between them. An *interlock* is a system that prevents an operator from performing a dangerous task. (Some cars use interlocks; for example they will not allow the engine to be started until the seat belts are fastened). A *trip* causes an automatic response to occur if a hazardous situation occurs (an example of this is an air bag in a car: if the car crashes, the trip system causes the bag to inflate.)

In plant terms, a trip causes some section of the unit to shut down suddenly in response to a potentially dangerous condition. A common example of this is a fire eye (flame detector) on a furnace. Loss of flame in a furnace can be very dangerous if the flow of fuel is not stopped. Therefore fire eyes are installed. If the fire eye detects a loss of flame, it immediately shuts off the flow of fuel to the burners. This is a trip.

Volume II should contain a description of all alarm, interlock and trip systems, what they do, what causes them to be activated and how to respond to them.

Material Safety Data Sheets (MSDS)

MSDS are described in the OSHA standard 1910.1200(g). They are not generally included in the procedures. Nevertheless, it is suggested that Volume II contain a list of the MSDS for the plant that is being operated and that it should identify where they can be found.

MSDS are produced and written by the manufacturer of the chemical (or its importer.) They are not usually produced by the user of the chemical. In practice, some companies may decide that the quality of the MSDS with which they were provided is not good enough, and they may decide to work with the manufacturer on making a better MSDS. Manufacturers, importers and distributors are responsible for ensuring that all users have an MSDS. The user must make sure that the MSDS is available to all affected employees.

The MSDS must provide information on all safety aspects of the chemical in question, including toxicity and fire hazards. It should also provide information on what actions are to be taken in the event of an accident occurring.

The manufacturer must provide information on the MSDS as to how he can be contacted in the event that emergency assistance is required.

Safety Equipment and Clothing

Many plant activities require the use of special safety equipment or clothing, in addition to standard clothing such as hard hats and safety glasses. It will be recalled from Chapter 2 that special safety instructions for particular operations are best included with the title blocks of the Volume II modules to which they apply. The safety section of Volume II provides background information on issues such as the brands of equipment authorized for use, means of testing gloves for leaks and the use of goggles and face shields.

Permit To Work Procedures

Permit-to-Work procedures describe the actions that the operators need to carry out before maintenance work, usually carried out by members of other departments, can start. They are explicitly called out in OSHA's PSM standard.

The basic idea behind most Permit to Work procedures is to keep operators and maintenance workers from contacting sources of high energy. Therefore, at the heart of most of these procedures is a description of how a high energy source is to be physically isolated, along with a description of the managerial and organizational mechanisms that ensure that the system works properly. Some types of high energy source are described below.

Physical Energy

Physical energy on a plant is present in many forms, such as rotating machinery, falling objects and vehicle movement. Physical energy can also take the form of potential energy, such as the energy contained within a high pressure system.

Chemical Energy

Chemical energy occurs in the form of high concentrations of toxic, reactive or corrosive materials. Workers can be exposed to it if there are problems with breaking flanges, entering vessels or handling drums and cylinders of hazardous chemicals. The Permit to Work procedures in this area focus on the *First Break Policy* (usually the unbolting of the first flange), *Line Blinding* (the blind provides a physical barrier between the chemical energy and the workers), and *Vessel Entry* (determining if a vessel is safe to enter).

Flammable Energy

One form of chemical energy that is particularly common on process plants is the energy associated with something catching fire (or exploding.) For this reason, most plants have very strict procedures concerning *Hot Work*, i.e. the use of open flames and spark-producing equipment.

Electrical Energy

Electrical energy is an insidious of high energy sources because it is not visible: it is not possible to tell simply by looking at an item as to whether or not it is energized. For this reason, *lockout/tagout procedures* tend to emphasize the problems associated with high voltage sources.

What are called *Permit to Work Procedures* in this book are given different names by different companies. Sometimes they are called *Standard Operating Procedures* or *Safe Work Procedures* (this is a phrase used by OSHA.) In addition to the PSM standard, the appropriate OSHA regulations fall under 29 CFR 1910.252. They cover topics [10] such as:

- Hot work (welding, cutting, open flames)
- Vessel entry (vessels, voids, closed spaces)
- Line break
- Disposal operations

- Excavations
- Work in the dark
- Truck unloading and loading and vehicle movement
- Handling of toxic materials
- Use of matches and lighters

Writing Permit to Work Procedures

Most companies have their permitting procedures in good order; if they did not there would almost certainly be an unacceptably high accident rate. This does not mean, however, that these procedures cannot be improved. Indeed, many of the techniques described in this book, such as modularization and the use of graphics, are equally valid for Permit to Work procedures.

One feature of Permit to Work Procedures is that, even more than normal operating procedures, there is tension between the need to be both rigid and flexible. On the one hand, they have to be very rigid because even a small deviation from procedure could lead to a major accident. Therefore, it is essential that they be written very tightly, and applied precisely. On the other hand, they have to flexible because they cover an almost indefinite number of potential maintenance scenarios.

It is particularly important to have Permit to Work procedures because they often cover an interface between two departments, usually operations and one of the maintenance groups. Therefore, the potential for accidents is increased because there are so many possibilities of communication slips, particularly if the maintenance group includes contract workers who are not familiar with the plant or the company culture.

ELEMENTS OF PERMIT TO WORK PROCEDURES

As with all types of operating procedure, the decision as to what is appropriate for a particular company and a particular plant site depends on the individual circumstances and requirements. This section merely provides some general guidelines.

Plan the Job

The first step in the procedure is for the operations and maintenance personnel to plan the work that is to be done. The operations staff need to make sure that they fully understand what impact the work will have on plant production, and how long the work is expected to take. The maintenance staff must be provided with an understanding of the general nature of the equipment and the process with which they will be working.

An important element of the job plan is to have some type of map or master card that shows all the isolation points associated with the job. For example, if a pump is to be taken out of service so that its impeller can be replaced, the operations and maintenance staff should have a drawing, possibly a fragment of a P & ID, that shows where blinds and other isolation devices are to be inserted. There should be equivalent drawings or sketches for other high energy systems, such as the electricity to the pump's motor and utilities such as plant air or nitrogen, if they are part of the pump assembly.

Lock Out the High Energy Sources

Once the job has been planned, and the operating conditions adjusted to compensate for the effects of the upcoming maintenance, the next important step is to isolate the energy sources. This usually involves a lock-out/tag-out system.

Using the example of the pump that is to be torn down, each of the valves in the lines to that pump are closed and then locked so that they cannot be opened inadvertently or without proper authority. The most secure system is to provide *every* worker on the unit with a personal lock, to which only he has the key (there needs to be a spare key in the event that the original is lost, but the backup key is not intended for normal use.) If the worker is assigned to this particular maintenance task, he locks the valve with his own lock. This means that the valve cannot be re-opened until he has removed the lock. Such a system minimizes the chance of putting equipment back in service before all the maintenance tasks have been properly completed.

Some companies use a lock-out box. The equipment itself is locked out with a master key that is then put in a special box. Each worker on the job then attaches his personal lock to that box so that it cannot be opened until they have all completed their assignments.

It is not always possible to physically lock out a piece of equipment. In such cases, the equipment should be tagged with a distinctive marker that has all the effect of a lock except that it would be possible for someone to physically over-ride it, should they choose to do so.

The locks should be properly *tagged out* with labels that describe the nature of the task, and provide the appropriate warning information.

Prepare the Equipment

Once the equipment has been locked and tagged, it has to be prepared for maintenance. In many cases, this means that the equipment has to be purged of hazardous and/or flammable chemicals, either with steam or with an inert gas, such as nitrogen. The Permit to Work procedures need to spell out how the operations staff will know when the equipment is ready for entry, or when it is safe to open a flange on a line.

First Break

Once the equipment has been isolated, and readied for maintenance, the *first break* takes place. This is a critical time because the process equipment is to be opened to the atmosphere and, if any mistakes have been made in isolating the source of high energy from the work site, this is the time that it is likely to reveal itself.

For this reason, procedures such as the first-break of a line, or removing the first manway from a vessel, should be carried out with the greatest care. The Permit to Work Procedures should require that at least two people should be present (the second should be a "buddy" located some distance from the work who can warn of problems should they occur.) This is also the time for maximum precautions concerning the use of safety clothing, breathing air and having emergency equipment on stand-by.

Continue the Preparation and Start the Work

Once the equipment has been locked out, and the first break has been made, the maintenance workers can proceed with their work. Caution should always be taken because of the possibility that there might be isolated pockets of chemical or flammable materials that were not successfully removed during the preparation steps. For this reason, the procedures should require, for example, that the oxygen content of a vessel which is being worked on remains at the correct concentration, and that no hazardous chemicals have had a chance to intrude.

The maintenance work itself can sometimes expose a high energy source. This sometimes happens when welders are working on corroded equipment, such as the linings of storage tanks. Their welding torch may burn into a pocket of chemical that may be located in the interstices of the metal plates.

Closing Down the Job

All the precautions concerning starting the job apply to closing it down and returning the equipment to operations. Therefore, it is critical that the Permit to Work system contain a provision for closing out the job. Just as blind lists, lockout and tagout procedures are needed to isolate sources of high energy from the maintenance workers, so a procedure is needed for making sure that all of the isolation devices are removed safely. A perennial problem, for example, after major turnarounds is that blinds are overlooked and left in place. If this happens, and the plant is re-started, it will then be necessary to go through this whole procedure of isolation one more time. This is embarrassing and will probably lead to economic losses. Even worse, there may be strong temptations to short cut the work the second time around, and someone may get hurt.

The importance of this step — proper job closure and hand-over to operations — cannot be over-emphasized. One of the root causes of the *Piper Alpha* accident (the North Sea oil platform that was destroyed by explosion and fire) was the fact that there were two open work orders on a compressor. The first work order was closed out and operations department restarted the compressor, not realizing that a second work order had not been closed out and that there was an open line adjacent to the compressor. Consequently, as soon as the compressor was started, large quantities of light hydrocarbons were released; they then ignited.

Design of the Work Permit

Integral to all of the above discussion is the system for issuing work permits. Some of the items to consider when designing or auditing such a system are listed below.

Authority. It has to be established who has authority for issuing permits, and for closing them out when the work is finished. Generally, this will be the senior operator in the area of concern, and his counter-part in the maintenance organization. If there is a shift change (either operations or maintenance) during the course of the work, this should be properly documented and controlled.

Scope of Work. It is essential that the work permit properly define the physical scope of work: the equipment to be worked on, the location of line breaks and the location of all blinds. This will include the lockout sketch described above.

The Permit Form. A permit form should contain the following items:

- Location of the job
- Equipment to be worked on
- Nature of the work
- Equipment and tools to be used
- Start date and time
- Finish date and time
- Special precautions
- Pre-work inspection
- Post-work inspection
- Signatures

OPERATING LIMITS

Developing and communicating operating limits is one of the most important requirements of OSHA's PSM regulation, and it is a foundation on which a process safety management program is based. It is also the foundation of a good operating manual. Deviations need to be defined quantitatively. Figure 19, which uses the three-part modular concept described in Chapter 2, is one way of communicating operating limits.

Figure 19

Example of Operating Limits

Instrument Tag	FRC-121 *Board Mounted*	**Units:** kg/h
Instrument Name	Flow Recorder, from Light Ends Surge Drum, V-100 to Light Ends Column, C-121	
Control Scheme	This is a dependent variable, whose set point is controlled by LIC-120.	
Purpose	Measures and controls the flow of recycle isobutane to the Light Ends Column	

Lo-Lo	**Lo**	**Normal**	**Hi**	**Hi-Hi**	**Alarm/ Interlock (Trip) Settings**	**Discussion**
-	10000	17000	25000	27000	25,000 (hi alarm) 30,000 (interlock)	If the high flow rate is reached, C-121 may flood.
Authority Box						

TITLE BLOCK

The Title Block describes the instrument whose range of values is provided in the Information Block..

Instrument Tag Number

In this case, the instrument tag number is FRC-121, where the letters FRC stand for Flow Recorder Controller. The signal and the control function are board mounted.

Units

This provides the units of measurement. In this example it is kg/h. The units should always be physically meaningful. An instrument output signal is unacceptable because, if the instrument is re-ranged, the information in the procedures will become misleading.

Instrument Name

The name of the instrument should be sufficiently detailed to give a good description of what its function is, its location, and what it is measuring.

Control Scheme

This section of Volume II provides information as to how the instrument fits into the overall control scheme. For example, if the signal from the instrument may used to directly control the position of a control valve (as is the case for FRC-121 in the Worked Example.) In more complex systems, the signal may set the set point for another instrument (cascade control.) The control scheme is shown on the P & ID.

Purpose

This section provides a brief overview of the purpose of the instrument.

INFORMATION BLOCK

The Information Block contains the columns shown in Figure 19.

Normal

The normal value for the variable is the target value during normal operations. There may be more than one value in this column if there is more than one way of operating the unit. If this is the case, there would be more than one row in the table. The normal flow in this case is 17,000 kg/h.

Lo and Hi

To the left of the normal value is the *lo* value. This represents the lower bound for normal operations. As long as the flow is above 10,000 kg/h, conditions are acceptable and no adjustments need be made in order to keep the operation safe. To the right of the normal value is the *high* limit. In this case, the value is 25,000 kg/h.

The range provided by the hi and the lo values does not mean that the operator is allowed to let the variable wander around anywhere within that range. In the above example, the operators daily instructions may tell them to run the flow at 16,500 kg/h with a range of ± 500 kg/h. The essential point is that supervision can make such decisions, and can issue such instructions, secure in the knowledge that they do not have to initiate a formal management of change to do so, and knowing that they are not jeopardizing safety.

The operators are not permitted to intentionally deviate outside the Lo-Hi range without obtaining permission, i.e. without initiating a management of change request and receiving the appropriate approval. In practice, particularly during an upset condition, an operator or a supervisor may take upon himself the responsibility of moving outside the approved range, but he should fully understand the implications of making such a decision. If the value does go outside this range, then the operator knows that corrective action must be taken.

The lo and hi values represent the "trouble", not the "emergency" boundary. Although it is often difficult to clearly distinguish between the two, they *are* different; it is important to understand the distinction because the appropriate response may vary dramatically. If the lo and the hi values merely represent "trouble", then the operator will not necessarily trigger an unsafe condition by deviating outside the range. But he will cause some type of economic upset.

Hi-Hi and Lo-Lo

In addition to the *hi* and *lo* values, *hi-hi* and *lo-lo* values can also be provided. These usually represent a true emergency condition. It is usually difficult to establish what operating limit values should be. This is particularly true with emergency limits because there is necessarily almost no practical experience of what they should be since emergency situations occur only rarely.

Alarm Values

This column contains any alarm values associated with the instrument. If the system is controlled by a DCS (Distributed Control System), there will usually be two types of alarm: "soft" and "hard." A *hard alarm* is one that has its own dedicated circuitry: it cannot be adjusted by the operators. If a hard alarm sounds, a serious emergency condition has been enunciated. *Soft alarms*, on the other hand, are included in the normal instrument wiring. Often the operators are free to adjust them, and they tend to warn of operational problems rather than true emergencies.

In the example, there is a hi alarm set at 25,000 kg/h. (Note that this alarm enunciates before the Hi-Hi value is reached.)

Trips and Interlocks

Trip and Interlock values can also be provided. In Figure 18, an interlock/trip is initiated at 30,000 kg/h. There is no low flow interlock.

Discussion

This final column also contains a general discussion of the variable and the operating conditions with which it is associated. For example, there may by a discussion concerning instrument reliability. In this example, there is a brief statement as to the anticipate consequences of operating above the stated alarm value.

EQUIPMENT AND VALVE INFORMATION

This section of Volume II provides information on the major items of equipment used in the process for which the procedures are being written.

EQUIPMENT

Volume II will usually contain a list of equipment on the unit. Usually, the development of this list is quite straightforward, particularly if the original data sheets are available and pertinent.

One problem to watch for with equipment lists, and indeed with all references to equipment, regardless of which manual they are in, is that the operators often use idiomatic terms or nick-names for their equipment, rather than the official name, as provided on the P & ID. For example, what is officially referred to as the "Light Ends Recycle Compressor" may actually be referred to as the "Waste Gas Compressor" by the operators. They will sometimes use the name of the manufacturing vendor as a nick-name. So, an item that is officially known as the "Reactor Feed Pump" may be referred to as "The Richardson" because that is the name of the company that manufactured the pump. The writers of the procedures have to decide whether to use the official name or the nick-name. The official name is correct and authorized, and is not open to interpretation. The nickname, however, is the one that is actually employed by the users of the manual: the operators.

Volume II should provide detail of checklists for the equipment that needs to be checked before a startup, and at regular intervals during normal operation. Examples of information of this type includes:

VALVES

Valve Positions

Volume II should contain a table showing the position for each block valve during normal operation and when the plant is shut down. The table can also describe the way in which the valve is secured during a turn-around (tagged, locked or chained.)

Relief Valves/Rupture Disks

There is not much that an operator can usually do to check a relief valve or rupture disk, except to make sure that it is not leaking and that there are not visible problems, such as a blind inserted between the relief valve and the process that it is protecting. The actual testing of its set pressure is usually carried out in the workshop.

Bleeders, Vents and Sample Points

It is important to provide a list of bleeders and vents. Many accidents occur when they are accidentally left open, or when a valve sticks open, or when a valve handle breaks off.

Blind Lists

Plant safety at times of maintenance and turnaround depends heavily on the use of blinds because they provide a positive isolation between two sections of the plant. Blinds are often placed in lines so that one section of the plant can be opened up and ventilated to allow maintenance workers to enter it. Clearly, an incorrect assumption that a blind has been installed creates a potentially very dangerous situation.

It is vital that a clear, easy-to-use blind system be created and that the operators and maintenance staff are properly trained in its application.

Features of the blind list system should include the following.

1. There must be a clear, accurate record at all times of what has been blinded.

2. It should be possible to locate a blind at a glance once it is installed in a flange. One very simple system is to create a set of metal tags, each of which is stamped with a unique number. If a line is to be

blinded, one of the tags is removed and attached to the blind. Then a record of what the blind is, its number and its location is kept in a central log book or on a bulletin board. Further sophistication could be obtained by showing the location of the blind on the control room P & IDs.

FORMAT OF VOLUME II

Volume II, unlike Volume I, can use a conventional page numbering scheme because it does not use a modular concept. Therefore, it does not require any special formatting. Nevertheless, it will require frequent updating, and must be designed with this requirement in mind. It is suggested that the page numbering sequence start afresh with each section of the manual.

One method of formatting sometimes used is a two column approach, with titles and subtitles in the outer column, and the associated text in the inner column. A layout such as this makes it easier to find information because some indexing is built in. This approach is made even easier if the key words are printed on the extreme right of the page in a shaded box.

CHAPTER 4
VOLUMES III AND IV — EMERGENCY & TROUBLESHOOTING PROCEDURES

INTRODUCTION

This chapter describes the writing of procedures for handling situations where one or more variables are no longer within the permitted operating range (see the discussion on operating limits in Chapter 3.) The consequence of a deviation may be quite minor, or it might have catastrophic potential.

There are four types of operating manual that will be discussed here. They are:

1. The Emergency Operations Manual. This describes how to keep the plant running during an emergency.

2. The Emergency Shutdown Manual. This describes how to shut down the unit quickly and safely. The Emergency Operations and Shutdown manuals are included in Volume III.

3. The Emergency Response Manual. This provides site-wide instructions to be followed when the initial emergency is out of control. The distinction between *emergency operations* and *emergency response* is made in OSHA's §1910.38 (a)(2)(ii).

4. The Troubleshooting Manual — Volume IV.

EMERGENCY SITUATIONS

During an emergency, error rates can rise to as high as 50% or more. These errors can include slips, mistakes and violations. When writing emergency procedures, it is important to keep the following factors in mind.

- *"Heroics"*. In many plants, there is a general rule that anyone who is not part of the Emergency Response Team (ERT) should always move away from a situation, not toward it. In practice, people are tempted to take "heroic" action — thus violating this rule. For example, if a pump seal starts leaking a hazardous chemical, an operator may choose to run up to the pump, shut it down and block it in rather than going toward the control room and announcing an emergency situation.

- *Buddy Loyalty*. Loyalty between individuals can lead to violation of rules. For example, if someone is knocked down by fumes, it is tempting for his buddy to go toward the situation and pull the person out without taking the proper precautions. This may result in having two people knocked down.

- *Fixation*. During an emergency, people are often overloaded with urgent information. Consequently, they may tend to "fixate" on just one or two items, even if they are a relatively unimportant part of the total situation.

- *Panic*. If a person panics, they are likely to omit actions that need to be performed.

TYPES OF ACCIDENT

EVENT-DEFINED ACCIDENTS

Many accidents start with a fairly well defined situation that tells the operators that an emergency may be developing. Such events include:

- *Failure of a major utility such as cooling water, electricity, instrument air or steam*. An event of this type will sometimes trigger a sequence of subsequent events that result in an emergency, although normally it is only likely to result in "trouble," i.e. it is not likely to have more than economic consequences.

- *Severe weather conditions.* A hard freeze may disable the cooling water system or block instrument air lines; heavy rain may cause pump motors to be flooded; very hot weather might overload the cooling tower, thus causing pressure relief valves in the unit to lift. Any of these conditions may be sufficient to initiate an emergency.

- *Loss of function of a piece of equipment.* If an item of equipment, such as a recycle compressor, fails, a chain of events that could cause an emergency may be initiated.

The key element in all event-defined accidents is that, if the initiating event is recognized and addressed early enough, an accident will not occur. Also, the events can mostly be predicted — there should not be many surprises in this area.

Result-Defined Accidents

Many emergencies first make themselves evident from their results; i.e. the operator does not know what caused the emergency to happen, all that he knows is that a situation is underway and that action is required. Result-defined accidents include:

- *Fires.* Most process plants handle large quantities of very flammable hydrocarbons. Therefore, fires are an obvious potential emergency.

- *Explosions.* Explosions and fires often go together. If a cloud of hydrocarbon gas is released, a substantial amount of time may elapse before it finds an ignition source which can cause it to explode. The explosion is then followed by a fire which is fed both from the original leak and from leaks created by the explosion.

- *Toxic Fumes.* Most chemicals handled by process plants are either toxic or else they can cause death by asphyxiation.

Emergency Planning

When planning for emergencies, consideration must be given to the fact that general conditions at the time of the accident are likely to be bad. For example, the emergency may well occur at night, in bad weather, at a time when the number of people available is less than normal. The operators' training needs to reflect such conditions.

An emergency is not a time for individual judgment to be called into play. Once the immediate crisis in past, and the plant has been brought to a safe and stable condition, then judgment and thinking can be included in the response. But, in the first few critical minutes, the response must be automatic. This means, therefore, that the operators should have the emergency response procedures memorized and rehearsed. An emergency is not a good time for reading and creative thinking.

The Emergency Operations and Shutdown Manual — Volume III

Volume III tells the operator how to operate and/or shutdown the unit during the first stages of an incipient emergency. It provides operating instructions for the particular unit in question and is to be used only by the operators on that unit. If the actions that the operators take fail to contain the emergency, the site-wide Emergency Response Plan will be initiated.

Emergency Operations

Emergency operating procedures describe how to keep a plant running during an emergency. This is often done when the emergency is taking place in another part of the facility. It is often better to keep a plant running than to shut it down. The act of shutting down can generate its own set of problems.

Emergency procedures are *operating* procedures. However, they differ from normal operating procedures in the following ways.

- Response usually has to be very quick.

- Some type of action *must* be taken; the option of doing nothing is not permitted. Even when the emergency interlocks and shutdown systems take care of the immediate crisis, the operators must still follow up correctly.

- The effectiveness of the response can truly be a life or death issue. It is also likely to have critical effects on the environmental impact and on the plant's overall economics.

- Because there is an almost infinitely wide range of accidents that might occur, the emergency procedures have to be flexible. Yet, because time is of the essence, and because an emergency environment does not encourage contemplation, the procedures need to be quite rigid, short and precise. This is a difficult balance to strike.

Most emergencies do not occur in isolation. Usually, there are a multitude of events going on at once. For example, a fire is not likely to occur spontaneously. One of the reasons for its occurrence may have been loss of power to a critical compressor. The same loss of power may also have lead to a loss of fire fighting capability and a deterioration in the quality of communications, thus magnifying the consequences of the crisis.

Even when an emergency is initially an isolated event, its occurrence may well initiate a train of subsequent emergencies. For example, a fire may burn through the main utilities header to the plant causing a simultaneous loss of cooling water, steam and power.

The problem of simultaneous failure is particularly serious in areas vulnerable to earthquakes. The earthquake that causes lines and vessels to rupture may also break the fire water header, thus disabling the critical safety system. This does not place the Emergency Response Team in an enviable position.

Emergency Shutdown

The emergency procedures will include a section on Crash Shutdown. This section tells the operator how to shut down the unit in a crisis without causing any further damage and without jeopardizing other units. In practice, a crash shutdown may not actually involve completely shutting down the plant; sometimes the safest operation is to leave it running in a total recycle mode. Indeed, during an emergency, the safest response may be to simply keep the unit running — a shutdown may only compound the number of problems to be addressed.

Format of Volume III

Volume III is a set of operating instructions. Therefore it can be formatted in the same manner as Volume I. The only major difference occurs in the way that it is used. During an emergency an operator does not have time to read through a large amount of information. In fact, the emergency procedures should be short enough to memorize. Figure 20 is an example of an emergency operating module. It uses the same format as other procedures, such as those shown in Chapter 2. However, the "Response/Discussion" column has been eliminated because action must be immediate, therefore there is no time to review or to discuss the actions and their implications.

It is sometimes convenient to prepare extremely brief summaries of the instructions in the emergency procedures, and to print this information on small laminated cards that the operators can carry around with them.

Figure 20

Emergency Response Module

MODULE NAME	*Emergency Shutdown of F-6301, Recycle Fired Heater*			
MODULE NUMBER	U.200.5.63	March 3, 1995		
	PERSON	ACTION	REV	√
1	#1 Operator	Sound the emergency alarm	0	
2	#2 Operator	Turn off the Burners	0	
3	#2 Operator	Block in the Feed at FCV-6303 (see picture)	0	
4	#2 Operator	Stop steam to: - 100-EX-11 - 100-EX-12 - 100-EX-25	0	
5	#2 Operator	Isolate the following tanks at the tank farm - T-101 - T-101A - T-369	0	
6	#1/#2 Operator	Follow normal shutdown procedures for F-6303 (Module U200.2.12).	0	

EMERGENCY RESPONSE MANUAL (ERM)

This manual contains the site-wide Emergency Response Plan, and describes how to handle a situation that is out of the control of the operators on the scene. For example, if there is a large fire, the emergency response plan describes how operators from other units can assist in controlling it. The plan also describes how the company works with outside agencies, such as fire departments and environmental authorities. This manual is not included in the sequence of operating volumes because it is not specific to a single unit.

Some of the differences between Emergency Operating Procedures and the Emergency Response Manual are shown below.

- A sizable emergency is likely to involve people who are not familiar with the particular process where the emergency has occurred. This includes people, such as external fire departments, who are not even employees of the company that owns the plant.

- There must be one person in the plant who is in charge of all emergency response activities. The normal chain of command will probably be bypassed until the emergency is over. The person in command of the plant during the emergency does not have to be a senior manager. The management skills required to run the plant on a day-to-day basis are different from those needed during an emergency. Therefore, the normal line management may choose to assign this responsibility to someone else, probably a shift supervisor or a unit superintendent. Reporting to this person will be an Emergency Response Team. The members of this team are trained in handling serious situations. They also have the correct equipment and protective clothing.

The purpose of the Emergency Response Manual (ERM) is to describe what actions to take during a site-wide emergency, and to explain what the personal responsibilities of all individuals are during the course of the emergency. Table 8 shows an abbreviated Table of Contents for a typical ERM. As in the case of the Emergency Operating Manual, it is suggested that a modular approach be used.

It can be seen from Table 8 that a considerable amount of reference material is included in the ERM. In principle, this information should be located in Volume II, the Operating Information Manual, but may it be best to have all emergency information in one place.

The ERM should make full use of graphics. For example, site maps should show the location of the fire monitors.

Table 8

Table of Contents of an Emergency Response Manual

I. Introduction

II. Emergency Operations
   A. Initiating an Alarm
   B. When to Shutdown
   C. Response to Alarm in Other Areas
   D. Use of Fire Monitors
   E. Use of Fire Extinguishers
   F. Evacuation Procedures
   G. Adverse Weather Conditions
   H. High Water/Floods

III. Equipment and Clothing

IV. External Coordination
   A. Other Plants In The Area
   B. Fire Department
   C. Police Department
   D. Regulatory Agencies

V. Loss of Utilities
   A. Loss of Cooling Water
   B. Loss of Steam
   C. Loss of Control Central
   D. Loss of Instrument Air
   E. Loss of Plant Air
   F. Loss of Nitrogen

VI. Emergency Shutdown
   A. Sequence of Actions
   B. Responsibility for Emergency Shutdown

Table 8 (continued)

VII. Equipment Failure
  A. Line Break
  B. Hose Break
  C. Excessive Flaring

VIII. Post-Emergency
  A. Cleanup following a Spill
  B. Notification of Agencies
  C. Decontamination

IX. Reference Material
  A. Alarm Codes
  B. Area Responsibilities
  C. Description of the Fire Water Header
  D. Telephone List
  E. Fire/Police/Ambulance Coordination
  F. Mutual Aid
  G. Emergency Organization
  H. Emergency Control Center
  I. Emergency Drills
  J. System Test and Maintenance
  K. Terrorist/Security Threats

Other OSHA Emergency Response Regulations

Some of the other OSHA regulations concerning emergency operations and response are listed in Table 9 (details of some of them are provided in Table 16 in Chapter 8.)

Table 9

Other OSHA Emergency Regulations

I. Employee Emergency Action Plans
   A. 38(a)
II. Ventilation
   A. 94(d)((9)vi)
   B. 94(d)(11)(v)
III. Emergency Planning and Response
   A. 119(n)
   B. 120(a)
   C. 120(b)(2)
   D. 120(f)(2)
   E. 120(q)(1)/(2)
IV. Eye and Face Protection
   A. 133(a)
V. Respiratory Protection
   A. 134(b)(7)
   B. 134(e)(1)
VI. Medical Services and First Aid
   A. 151
VII. Fire Brigades
   A. 156(b)
   B. 156(d)
   C. 156(e)
   D. 156(f)
VIII. Portable Fire Extinguishers
   A. 157(b)
   B. 157(d)
   C. 157(g)
IX. Fixed Extinguishing Systems
   A. 160(c)
X. Fire Detection Systems
   A. 164(e)(3)

Table 9 (continued)

XI. Employee Alarm Systems
- A. 165(a)/(b)/(c)
- B. 252(f)(13)

XII. Material Safety Data Sheets
- A. 1200(g)

TROUBLESHOOTING — VOLUME IV

Troubleshooting covers situations that are not critical; issues such as equipment reliability and product quality fall into this category. Troubleshooting guidance is provided in Volume IV.

THE NATURE OF TROUBLESHOOTING

Troubleshooting has been defined [11] as "the search for the hidden cause or causes that leads to inadequate performance." A troubleshooting response is appropriate when there is no anticipated danger to personnel and when there is little chance of any significant equipment damage; usually the main concerns are about product quality, production rates, productivity and equipment repair costs. Table 10 shows some of the differences between emergency response and troubleshooting.

Table 10

Troubleshooting and Emergency Procedures

Emergency Procedures	**Troubleshooting Procedures**
Speed is essential — time is of the essence.	Usually there is time to review what needs to be done.
The consequences of incorrect response are likely to be catastrophic.	The consequences of incorrect response are primarily economic.
The instructions must be short in number and in length.	The instructions can be discursive. Different points of view can be presented because there may be different causes that generate the same symptoms and because more than one solution may be viable.
The instructions must be followed as written.	The instructions are closer to being guidance or suggestions; there is no absolutely correct or incorrect way of addressing the situation.

In most troubleshooting situations, time is not of the essence; management and the operators have time to consider what is the best action to take. Typical situations of this type include the protective measures that have to be taken to anticipated harsh environmental conditions such as a hard freeze, storms and floods. Troubleshooting is also invoked for quality problems, such as changes in feedstock quality or market composition.

Ideally, troubleshooting measures are implemented before the problem becomes too serious. For example, if experience shows that a certain product line often leads to the breakdown of some equipment items, the measures to prevent this from happening should be taken before the manufacture of the new product commences.

One of the most important decisions that an operator has to make when an abnormal operation occurs is to determine whether or not the situation with which he is faced represents "trouble" or an "emergency." This decision often has to be made quickly and under considerable pressure. A good set of operating procedures gives as much help as possible in this situation, i.e. assistance on how to diagnose what has gone wrong, and what the immediate response ought to be. The target conditions and acceptable operating ranges for each major variable that are provided in Volume II are very important here. As much as possible, they should provide the operator with a quantitative basis for his decision.

If the operator were to handle an emergency as if it were only an operational problem, the consequences could, of course, be very serious. However, the opposite situation can be almost as bad. If he erroneously decides that *trouble* is an *emergency*, not only will his subsequent actions lead to unnecessary production and productivity losses, they may actually create an emergency. For example, tripping a motor-operated valve to its closed position in response to a mis-diagnosis of an emergency can cause hydraulic and thermal shocks in other parts of the process that might cause an upstream flange to spread, leading to a possible release of toxic or flammable chemicals. Even if the consequences of the action do not lead to such extreme consequences, the sudden shutdown will likely cause irreversible degradation. For example, catalyst activity may be permanently reduced, equipment will be physically stressed, and the operators themselves will be tired and probably more prone to making mistakes in the time immediately following the false emergency.

Some of the distinctive elements of troubleshooting procedures, and the way in which they differ from normal operating procedures and emergency procedures are described below.

- Normal operating procedures are "closed," that is they fully describe a particular operation. Troubleshooting, on the other hand, is an open-ended activity. There is an almost infinite number of potential problems. Therefore the manual can never be comprehensive or definitive; all it that it can do is to offer guidance in the form of suggestions and ideas based on experience and brain-storming (such as would come from a process hazards analysis.) There is a wide range of possible actions for any troubleshooting situation; many of the actions will be new and untried.

- Development of a troubleshooting response often requires insight and imagination because there is rarely any pre-defined "correct" response: each situation possesses unique qualities. This aspect of troubleshooting is one of the reasons for stressing the education, as opposed to training, of the senior operators. Training teaches them how to do one particular task, education teaches them the principles on which the plant is running and how to think through problems.

- Troubleshooting often involves differing opinions, many of which will conflict with one another, and none of which are necessarily absolutely right or wrong because different causes may result in the same symptoms.

- Troubleshooting procedures are organic. (This is true of all procedures, but particularly those for troubleshooting.) As more and more operating experience comes off the plant, so the amount of useful troubleshooting material will increase. Their usefulness depends on the commitment that plant personnel make toward recording their experiences whenever a problem arises.

- There is an element of creativity in troubleshooting that must be rigorously excluded from normal and emergency procedures. In particular, when solving problems, an operator will often reason by analogy, i.e. he compares the present situation with a similar, but not identical, situation that occurred at some other time and place. He then has to decide if the analogy is appropriate, and what action should be taken as a result of this reasoning. There is no place for this type of approach in the emergency response procedures.

- As expert systems mature, troubleshooting will be an important application area.

Troubleshooting Using The HAZOP Technique

Troubleshooting can be considered to be a three-step process. It is described below, and is illustrated in Figure 21 which is based on the Hazard and Operability (HAZOP) method discussed in detail in Chapter 8. It is also based on the Operating Limits table (Chapter 2.)

Step 1 — Identify Causes of the Problem

The first step, is to identify the hazard or problem, i.e. the trouble that is to be "shot." Sometimes this is obvious. Problems such as mechanical failures, leaks, off-spec lab results are all easy to spot. However, the process of hazard identification can be considerably more subtle. For example, the temperature in a reactor bed may go up slightly for the following reasons:

1. Catalyst poisoning
2. Reduced reflux rates due to heat exchanger fouling
3. Erroneous lab analyses

If quantified target values and ranges for key variables are available troubleshooting becomes much easier because it reduces the need for individual judgment as to whether there is a problem at all. It also greatly assists in determining whether or not the deviation represents operational trouble, or the beginnings of an emergency.

Step 2 — Determine the Consequences of the Problem

Once the problem has been identified, and it has been established that it is not an immediate emergency, the operator has to determine possible safety, environmental and operability consequences. He also has to decide how much time he has to respond. If there is sufficient time, he should consult with his management and colleagues before making any moves.

Step 3 — Provide Some Suggested Responses

The final step in the troubleshooting guidance is to provide the operator with suggestions as to how he can respond. It is important to understand that, in most cases, this will be no more than guidance; it is *not* a set of instructions. The operator should be encouraged to think creatively in non-emergency, unfamiliar situations.

Figure 21

Troubleshooting Guide

From	V-100, LIGHT ENDS SURGE DRUM			
To	**C-121, LIGHT ENDS COLUMN**			
	Value	**Possible Causes**	**Consequences**	**Possible Responses**
Flow **FRC-204**	Normal Range: 1700 - 2500 lb/h			
Hi	> 2500	There is no operational experience of this having happened. The flow to C-1211 is limited by pump capacity and line size.	Flooding of C-1211: primarily a product quality problem	
Lo/No	< 800	1) Feed pump, P-123A or B, stops. 2) Control valve, FCV-1219, fails closed. 3) Polymer build-up in the feed line	Reduction in plant capacity; product quality problems	If the cause is mechanical, there is usually about 20 minutes of operational time available before action has to be taken. If the cause is polymer build up, inform supervision and be ready to switch this stream to the blowdown tank, V-9200.

Figure 21 (continued)

Reverse		During startup, pressure in C-1211 may be greater than V-100. There is no check valve on P-123A/B.	Operational. It may cause catalyst to enter V-100. This is not hazardous, but it can lead to a severe economic loss.	Watch the pressure in C-1211 closely during startup. If reverse flow is suspected, block in the feed line to C-121 until the problem is resolved.
Wrong Type Of	Polymer build up	See low flow		
***Temp* TRC-1219**	Normal range: 230 -260°F			
Hi	> 260	Fouling of the reflux cooler, E-122.	Product quality problems and possible rate restrictions.	If the temperature exceeds 285°F, shut down immediately. (Reference Emergency Operations module 129.4.12.)
Lo	< 230	In very cold weather, the viscosity of the stream will rise.	Operational. It can cause P-123 to trip out on high amps.	Make sure that the outlet temperature from E-214 is at least 250°F.
***Press* PI-1205**	Normal range: 75 -120 psig			
Hi	> 120	High pressure can occur if the inlet pumps are dead-headed.	See low flow	
Lo		No issues identified.		

- The plant is divided into physical sections (these are roughly equivalent to the nodes of a HAZOP.) Each section represents a process step in the unit, i.e. there is normally a major change in one of the key variables at the node. In this example, the node is the line from V-100 to C-1211, through P-100. The process change is the increase in pressure provided by P-100. A convenient way of creating sections is to select the more important instruments and instrument loops.

- A matrix of process and deviation keywords is set up. The standard process words are "flow," "temperature" and "pressure." It may also be useful to put in a "miscellaneous" keyword. The possible variations on each of these keywords is provided. Generally they are: "hi," "lo," "reverse" and "wrong type of."

- There are five columns in Figure 20. The second of these provides a quantitative value (where known) for target conditions and for deviations. The remaining three columns for the matrix are: "causes," "consequences" and "possible response/suggested action."

- Many of the cells — particularly those for the possible responses — will remain blank, indicating that no troubleshooting experience is available for in these cases.

- The Possible Response column provides guidance to the operators. It also serves as a check list.

USE OF EXPERT SYSTEMS FOR TROUBLESHOOTING

Nature of Expert Systems

People react in different ways at different times when they are handling plant upsets. Sometimes they reason through a situation using knowledge and logic. At other times, they simply follow a set of rules or guidelines in a routine way. And sometimes they simply compare the present situation with one that they have experienced, either directly or by anecdote, and then determine whether or not the two situations are similar enough that the lessons learned from the first incident are applicable to the present incident. In other words, they think by analogy. It is this way of thinking that is mostly used when troubleshooting plant problems.

Computer programs that are intended to mimic expert knowledge and experience are known as expert systems. If such a program has sufficient information, and if that information is properly organized, it will play the role of an on-line expert or troubleshooter. Expert systems, like any other type of computer program, work using rigorous logic — at its heart it is basically a set of ON/OFF switches. People, by contrast, do not usually think in such a logical manner; they tend to think by analogy and to compare present situations with similar, but not identical, situations that happened to them in the past.

Some of the key ideas behind expert systems are as follows.

- Most plant problems are not original — it is likely that another person on another shift has already experienced a situation similar to the particular problem that is being addressed. It is also likely that that person will have good idea of how to solve the problem. If this experience is captured by an expert system, it is available for everyone to use whenever it is needed.

- In any plant, there is a steady loss of valuable experience, as senior operators and engineers are transferred or retire (this trend was particularly noticeable in the late 1980s.) Before these people leave, their experience can be captured by the expert system. Eventually, given enough time, the expert system can in principle, become more knowledgeable than any single individual.

- Operators can practice responding to different problems by running them on the computer. This is particularly valuable training for how to handle rare, but important occurrences, such as fires or explosions.

- Expert systems are not forgetful, i.e. they do not make slips, nor do they suffer from fixation. Therefore, they have the potential to serve as an excellent check list for, say, determining which of twenty possible heat exchangers might be leaking internally. Similarly, the reasoning of an expert system is not clouded by emotion, competitive spirit, prejudice or feelings.

- A good expert system will stimulate discussion and disagreements among the operations staff. The results of these discussions can then be incorporated into the expert system — which is constantly growing and being modified.

Structure of an Expert System

Expert systems contain two parts:

1. The shell which organize the logic of the question and answer process, and

2. The information obtained from the experts.

This structure is analogous to the program/data organization of normal software. However, it is not exactly the same. In particular, the expert system can place confidence factors on the information and it can generate more than one feasible solution to one problem.

Limitations

Although expert systems offer great potential in this area, they have yet to become widespread. There are a number of reasons for this.

First, in order to be useful, an expert system must be large. Many examples tend to be trivial in size and are not useful in a real-world situation. Yet, the effort required to develop a truly useful troubleshooting expert system can be enormous, one that very few companies have ventured to undertake, at least in the area of process plant troubleshooting.

A second, more subtle problem with expert systems stems from the fact that computer logic does not really mimic the way in which people really think. As already discussed, when addressing a problem that requires troubleshooting people do not think entirely logically, instead they think creatively and they think by analogy. When attempting to solve a problem, people tend to think of similar, but not identical, situations that occurred in the past, and with which they compare the present situation. Arguing by analogy in this manner is the antithesis of how computers work [12,13]. Neither approach — strict logic or reasoning by analogy — is inherently superior; but they are fundamentally different. There will be a loss of confidence in the expert system if people do not see it mirroring their own thought processes. In particular, expert systems cannot jump to conclusions in the way that people do, i.e. they do not arrive at a logically incorrect answer that still provides useful insights in how to solve a problem.

A third limitation is that expert systems are unable to allow for human irrationality, usually expressed in some form of emotion. An expert system cannot know that Operator Smith is not on speaking terms with Operator Jones; it cannot know that management is very touchy about certain types of environmental release; it cannot predict the effect of emotions such as fear and excitement during an upset. Yet all of these emotional factors can effect on the way that humans handle a troubleshooting situation.

WORST CASE SCENARIO

One of the requirements of the EPA's Risk Management Program (RMP) is that companies prepare a "Worst Case" scenario. At the time of writing (early 1995,) agreement had not been reached as to what constituted such a scenario. The absolute worst case is that all of the flammable and toxic materials in the unit are released instantaneously, and that all restraints, *including passive restraints,* are ineffective. So, for example, if a company has a large tank of a toxic chemical, they have to prepare a scenario in which the tank suddenly and completely fails, and where all mitigation measures, such as earthen bund walls, are suddenly "not there." Except for the case of sabotage, such scenarios are so unlikely as to be incredible. For this reason, the EPA is continuing to evaluate just what the "worst case" should be.

In practice, the worst case for most companies will be the release of toxic chemicals into the surrounding community. Fires and explosions can be very dangerous to the plant personnel, but they rarely seriously affect members of the general public.

Basic risk management divides any scenario into two elements: consequences and probability. Usually there is an inverse ratio between them. In other words, there may be a chance of say 1 in 100 years that someone will be temporarily overcome by fumes, but the chance of large numbers of people being killed might be 1 in 10,000 years.

Risk can be expressed by the following equation:

$$\text{Risk} = \text{Consequence} * \text{Probability} \qquad (1)$$

In practice, a simple linear relationship as shown in Equation (1) is misleading because risk is fundamentally subjective. Risk is not an objective number in an engineering sense because it is *perceived.* So, for example, most people have a disproportionate fear of large accidents. Statistics may show that in a typical large city, something like 250 people a year die in automobile accidents. Many countermeasures are put in place, but the fact remains that people find this fatality rate more or less acceptable as being part of the cost of living in modern society. Yet, if once a year a jumbo jet carrying 250 people were to crash at that city's airport and kill all the passengers, there would be an outcry. In both cases the risk is the same: 250 fatalities per year. The difference is subjective: the airplane fatalities all occur simultaneously, they are not spread out over the year.

This being the case, the above equation should be modified as shown in Equation (2).

$$\text{Risk} = \text{Consequence}^{n} * \text{Probability} \qquad (2)$$

where n has a power greater than one. The value of n is inherently subjective. A figure of about 1.2 is sometimes used.

Using Equation (2) for the example of the airplane/road accident rate, it can be seen that, if a factor of 1.2 is used, the *perceived* risk of the airplane fatalities is greater than that of the road fatalities by a factor of $250^{1.2}/250$, i.e. about three times.

This discussion explains many of the difficulties that the nuclear power industry has had since its inception. The ultimate worst case is extremely bad. Yet the industry has responded by reducing the *probability* of an accident, not the consequence. This does not satisfy the public's concern.

From the point of view of the operating manual, it may become necessary to put the worst case situation into Volume III as a special operating procedure. The advantage of doing this is that it provides everyone, both inside and outside the facility, with instructions for what to do if the worst happens. The drawback is that it may be exaggerated and taken out of context by opponents of the company.

CHAPTER 5
VOLUME V — THE INITIAL START-UP MANUAL

INTRODUCTION

The first four chapters of this book have focused on the writing of operating procedures for plants that are already in operation. Therefore, knowledge as to how the plant works already exists. Consequently, the person writing the procedures does not need to create the operating information — he or she merely has to interview the operators and other process experts, and then to write down what was said in a clear, well-organized manner. This chapter discusses the writing of procedures for brand-new plants.

If the plant for which the procedures are being written is brand-new, this approach cannot be used because the necessary operating experience is not available. Information from other, similar plants can be used as a "go-by," and many of the operators and supervisors will have good, general knowledge of how to run plants. But specific knowledge concerning the unit in question will not be available until the unit is commissioned and lined out.

Because these instructions are used only on the first-time start-up, and possibly following major turnarounds, they should be put in their own binder, referred to here as Volume V — The Initial Start-Up Manual.

Throughout this book, the importance of clearly distinguishing between different procedures and information has been stressed. This distinction is particularly important in the case of Volume V.

The period prior to a first-time start-up is a very busy time. Consequently, there is a danger that the writing of Volume V will be deferred or skimped because it is not considered to be an absolutely essential part of the start-up — it is always possible to get the unit up and running, even if the manual has not been written. For this reason, it is very important to make sure that the writing of the manual is an integral part of the start-up plan. It is equally important to ensure that someone is responsible for making sure that the manual reflects last-minute equipment and operating changes.

Most Initial Start-Up Manuals incorporate information from a wide variety of sources, such as licensors, vendors and operators. Organizing all of this information into one coherent volume is a difficult challenge, particularly if some of the material has to be used without any type of modification or reformatting for legal reasons.

Throughout this chapter, a variety of check lists and action items is provided. It is important to understand that these are for guidance only. Actual manuals will need to match actual plant conditions.

DEVELOPMENT OF VOLUME V

One of the most distinctive features of the manual for a new plant is that it is written by different parties at different stages in the project. Typically, there are three groups:

1. The licensor of the process
2. The engineering company responsible for construction and commissioning (also equipment vendors)
3. The operating company

The role of each of these parties is described below.

THE LICENSOR

Many new plants use proprietary technology that is provided by a licensor. Usually, as a part of the licensing agreement, the licensor has to provide operating instructions. The operating company must follow these instructions — in particular, they must not allow process conditions to deviate outside the specified ranges. If they do not follow the instructions, the licensor does not have to assume responsibility for any problems that may arise. Any change in the licensor's wording, however minor, may lead to a reason for the licensor disclaiming responsibility for any subsequent problems, and it may void his warranties. There is no easy resolution to this dilemma. One way in which it can be addressed is to send the final manual to the licensor to ask for a review and clearance.

The licensor's operating instructions are rarely suitable for immediate use for the following reasons.

1. The instructions are usually generic. They do not tell the operators how to operate *their* particular plant. Some manuals of this type are so generic that they contain information on entire plant sections that do not even exist in the one being built.

2. Licensor manuals contain little or no plant detail. They are usually at Level 1 of detail (see Chapter 2.) They tend to be oriented more toward information than instructions.

3. Licensors only write procedures for their part of the process. Frequently this covers the reaction system, which uses proprietary techniques, but excludes the downstream product purification steps.

The Engineering Contractor

The detailed engineering contractor usually writes the first draft of the plant-specific manual. This should not be done until the final P & IDs have been issued. The format that is used is either specified by the operating company, or else it follows a style that was successfully used on earlier projects.

Problems that the contractor might face include the following:

1. He may be required to write the manual to fit the format and standard of his client: the operating company. This can be a problem if the contractor feels that the client's standards are inappropriate for the project in question, or if there is a conflict with the licensor's standards.

2. The level of detail that the client expects is often not properly defined at the start of the project. Nor is it discussed thoroughly enough during contract negotiations.

3. The contractor's personnel are usually technically-oriented people who are trained to design and construct plants. The skills that make them good at this do not necessarily make them good at writing manuals. In addition, they may not have much actual operating experience. Therefore, operating manuals written by engineering companies often emphasize the informational aspects (Volume II) but are relatively light on the actual operating procedures (Volume I.)

4. The engineering contractor may not know the details regarding equipment and instrumentation because these are still being specified and put out for bid while the manual is being written. The integration of vendor material with the manual may have to be done later.

5. Frequently, the plant is to be built in a different nation from that of the contractor. Therefore the manual will first have to be translated. This makes the use of simple, clear English particularly important

It is extremely useful if the engineering company has a well-defined standard that can be applied to any project, and if the company has people who specialize in the writing of procedures on their staff. It is also very important to define just what the manual should and should not contain at the start of the project. The resulting *Writer's Guide* (see Chapter 7) that would be included within their overall standards and specifications.

The Operating Company

The Plant Owner/Operator may take the information and instructions provided by the licensor(s), contractors and vendors and develop his own manual. He usually incorporates the following into the final manual.

1. *Company practices and standards.* Each company will have its own procedures and policies for running their plants efficiently and safely. In addition, the final manual will correspond with the way in which the operating company is organized. For example, the procedures for a plant where operations and maintenance personnel work on one another's jobs will be different from a plant where there is rigid job demarcation.

2. *Training.* It is likely that the first use of the manual will be to train the operators who have been hired on or transferred to the unit that is presently under construction. Therefore, the operating company is likely to design the manual so as to meet training needs.

3. *Vendor Information.* If the engineering company was not able to include all of the detailed vendor information in the manual, this will have to be done by the Operating Company.

It is enormously helpful if the Plant Operator can supply the engineering contractor with a well-thought out format and Writer's Guide *at the start of the project.* This can go a long way towards reducing the inefficiencies that can take place due to misunderstandings between the various parties as to what the manual is meant to be.

In Chapter 8, it is suggested that integrating a Process Hazards Analysis (normally a HAZOP) with the writing of the manual can be very effective because the activities assist one another. The HAZOP will often occur after the engineering contractor has finished his work. Therefore, it may be the client who will incorporate the changes to the manual suggested by the HAZOP team.

Ideally, the manual would be updated and re-issued within a few weeks of the start-up. This phase would incorporate the lessons learned from actual operation and would therefore reflect the reality of how the plant actually operates. This "as-operated" is analogous to the issuance of "as-built" P & IDs immediately prior to the start-up. In practice, this is the phase that often does not get done; its omission is the first step in the process whereby a manual becomes progressively out-of-date until eventually it is not used at all.

THE START-UP PLAN

The process of starting up a brand-new unit is commonly divided into the five major phases, as shown below and in Figure 22.

1. Mechanical Completion
2. Precommissioning
3. Commissioning
4. Start-Up
5. Line-Out

There is no universally agreed-upon definition of these terms (although a standard for mechanical completion and contractor/owner responsibilities is provided by API.) What is shown here is representative.

Figure 22

Phases of Start-Up

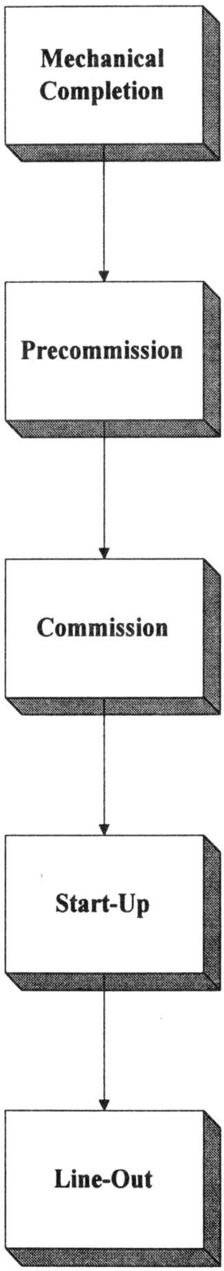

PHASE 1 — MECHANICAL COMPLETION

A unit is said to be mechanically complete when all equipment and instrument items have been properly installed. It marks the transition of control from construction to operations. At this point, operations will carry out the Prestartup Safety Review (discussed in Chapter 8.)

PHASE 2 — PRECOMMISSIONING

The purpose of precommissioning is to ensure that the plant that has just been transferred from construction to operations is safe and is built according to design. Equipment can be adjusted and modified. (Detailed mechanical completion and precommissioning instructions are provided by the API [14].) There are no process chemicals in the unit during precommissioning.

Some of the utility systems, such as the flare header or plant nitrogen, may be commissioned at this time.

Some typical precommissioning instructions, that can be incorporated into Volume V are provided here. (The details will vary from plant to plant, as will the order in which the instructions are carried out.)

1. Prepare Turnover Packages
2. Develop a Blind List
3. Develop a Punch List
4. Check Equipment
5. Check Piping Systems
6. Check Instruments/Instrument Loops
7. Check Safety Systems
8. Hydrotest

1. PREPARE TURNOVER PACKAGES

As the plant is being readied for commissioning, turnover packages can be prepared. The plant is divided into relatively small sections that share a common function. For example, a utility system, such as cooling water may be one package. A reactor system may be another. When the engineering company transmits the turnover package they are stating that *everything* to do with that system is now mechanically ready: all instruments are connected and working, all equipment is insulated and painted and all lines have been blown clear.

These packages have a contractual importance because it is at this point that control and custody of the unit passes from the engineering company to the operating company. The operating company will then organize the hydrotesting and commissioning.

2. Develop A Blind List

The blind shows where blinds can be located, where they are and what type should be used. This list must always be kept current

3. Develop A Punch List

The plant must be thoroughly punched out. This means that every item on the P & IDs and instrument loop drawings has been checked in the field. This is tedious and time-consuming work, but it is very important. When marking up drawings and lists, the following color codes are frequently used.

- Yellow means that the item checked out satisfactorily
- Blue is used to indicate that something should be removed
- Red means that additional material is to be added.

4. Check Equipment

All equipment must be installed and must have been inspected as compared with the vessel drawings and with the P & IDs. This includes the following for each item.

Vessels

- Temperature and pressure rating

- Metallurgy
- Elevation
- Lining
- Trays: number, type, spacing, levelness, liquid seal, weir height, number of valves and their operation, metallurgy, weep holes, and supports
- Distributors: location, alignment, size and number of openings
- Nozzles
- Vortex breakers
- Demisters: size, materials and support
- Thermowells: location and length
- Instrument connections
- Fireproofing
- Inspection plates in place
- Painting complete

Heat Exchangers

Prior to placing the shell and tube exchangers in operations, complete the following inspections.

- Check that the shell side and tube side services conform to P&ID requirement.
- Check location and orientation of nozzles against exchanger drawings and P&ID.
- Check exchanger sliding plate for free movement due to thermal expansion of the shell.

Pumps

Running in pumps is normally a two-part process:

1. Check the driver under no-load conditions
2. Check the operation of the pump and driver together

1. Driver Check- No Load

- Run the driver for a minimum of four hours uncoupled
- Check that the driver is running well

- Check control systems and over-speed controls

2. Pump and Driver Check

- Check rotation of electric motor drivers uncoupled from the pump.
- Connect the pump to the drive
- Check the pump and driver are properly aligned
- Check that overall installation is complete.
- Check water piping. Verify that water piping is connected, where required, to bearing jackets, pedestals, stuffing boxes.
- Check seal or gland oil piping. Conventional packed pumps in hot service are generally furnished with gland oil; verify this installation is correct and complete. When pump is furnished with a mechanical seal, verify that all of the components of the flushing system, such as strainers, separators, restriction orifice, and coolers where required, have been correctly installed and are clean.
- Verify that packing or seals are installed.
- Check that temporary strainers in suction piping have been installed.
- Verify bearings and shafts have been cleaned prior to final lubrication.
- Review the pump manufacturer's special instructions
- Rotate pump and driver by hand, verifying that they roll freely.
- Open the suction valves fully.
- Establish cooling water where required.
- Check that lubrication is satisfactory.
- If the pump is designed to handle a fluid with a specific gravity of less than 0.6, the use of water during run-in will probably over-load the motor. In these cases, a lighter, hydrocarbon liquid may be used.

5. Check Piping Systems

Check the following:

Piping

- Routing against P&ID. This includes all valves, line sizes and line specs.
- Bolts are in place and tight.

- Direction of flow for check valves.
- Insulation and steam tracing are properly installed.
- Line and equipment supports are properly anchored.
- Expansion loops are free to expand.
- Tie-ins are correctly installed.
- Blinds are installed where specified.
- Blinds are removed when required.
- Flanges are tight.
- Proper gaskets are used.
- Thermowells, sample points, vents, drains are provided when required.
- Plugs are tight.
- Low-point drains and high-point vents are provided.
- Valves operate freely, are accessible, and can be operated safely.
- Heater burners and internals are installed.

Nozzles

- Orientation of all equipment.
- Internal drawoff piping.
- The location of process nozzles.
- The location of thermowells and thermowell injections into vessels.
- All vortex breakers and risers.

Distributors

- The orientation and location against vessel drawings and the P&ID.
- Drilled holes and/or spray nozzles against vessel drawings for size, number, and orientation.
- General cleanliness.

Vessels that will remain idle for long periods after hydrostatic testing should be purged with nitrogen and left with a slight positive pressure to prevent rust formation. This is required if inspection was completed before hydrostatic testing.

6. CHECK INSTRUMENTS/INSTRUMENT LOOPS

Service and Calibration

Preparation of the plant instrumentation for startup and operation should include the following inspections and tests.

- Visually inspect instruments to verify compliance with specifications
- Visually inspect instrument installation to very compliance with specifications
- Check the transmitter and receiver readings of each instrument.
- Pressure test instrument piping
- Test and calibrate instruments
- Test instrument air signal lines
- Test thermocouple wiring
- Final preparation

Pay attention to instrument range, pressure rating, connections, materials of construction, size, electrical characteristics, and special features.

Inspect orifice plates as soon as possible so that they will be ready for installation when line flushing is complete. This includes inspection of size, materials of construction, location of weep hole (if any), counterbore, overall condition, measurement of orifice bore and verification of information stamped on the tab.

Loops

All instrument loop checks must have been completed and preliminary instrument calibration carried out. All orifice plates and other instrument items must be installed. Check the following.

- All transmission systems of each instrument after installation.
- Sealing fluid, where required, in all instruments.
- Air system to all control valves. The control valves are removed while line flushing and this section of the air system must be re-tested when the valves are reinstalled.

- Control valve operability, including stem travel and action on air failure.
- Installation of meter factor cards.
- Operability of the "no break" emergency power supply for instruments.
- Location and range of level instrument internal float, external type displacement tube cages, or differential pressure type nozzles.
- Calibration of all pressure and temperature gauges.
- All instrument lead lines, both air and electronic.
- Instruments which must be read while manually adjusting valves should be visible from the valve location.

7. Check Safety Systems

The following should be checked for completion and operability.

- Firewater header pressure.
- Fire monitors.
- All hand fire-extinguishers are in place and are properly charged.
- The safety showers and eyebaths work, and have been flushed until the discharge water from them is clean.
- All foam and other fire-fighting equipment is in place and operable
- Safety equipment, such as safety clothing, PPE (Personal Protective Equipment) and first-aid kits is available.
- The flare system (if used) should have been flushed, cleaned, hydrotested and dried. Its oxygen content should be less than 1% (this operation may be left until later depending on the particular operating circumstances.)
- All pressure safety valves must be bench-tested and set before final installation. This testing and setting may be done while the plant hydrostatic testing is in progress.
- The emergency shutdown system.

8. Hydrotest

Prior to the initial start-up, the entire unit is hydrostatically tested to ensure the integrity of the welding and to confirm that the vessels, exchangers, piping and other equipment meet the design pressure specification. Any equipment, whether piping, vessels or exchangers, which has been modified, repaired or has had connections made to it by welding during a downtime or construction period preceding a startup, must also be hydrostatically tested.

Sometimes, an air pressure test is carried out before the hydrostatic test to identify any open lines or flanges.

Preparation

The test is conducted by filling the items to be tested with water and then pressuring it to test pressure. If the system is not leaking, the pressure should hold once the hydraulic pump is turned off. If the pressure starts to fall, the leaking point (which is often a gasket) can often be identified by the water flow. Different parts of the unit are tested at different pressures, using the turnover packages already discussed.

At least two pressure indicating gauges should be connected to the equipment or the system that is under test. One of the gauges must be a compound pressure-vacuum gauge located near or at the high point of the system. If the operator controlling the applied pressure cannot easily read one of these gauges, an additional gauge must be provided where it will be visible to the operator during the duration of the test. When testing vertical vessel systems, the test gauge is located at the bottom of the vessel because this is where the hydrostatic pressure is the highest.

A recording gauge is used in addition to the indicating gauges for tests on all major vessels or equipment. Charts from the recorder must be witnessed by those responsible for the tests, and included in the permanent inspection record.

Test and calibrate all indicating and recording gauges through the test range prior to the test.

Precautions

- Blind (or block and bleed) all safety valves, instruments and pressure gauges that could be over-ranged by the hydrostatic test pressure.

- Place fine mesh screens in the suction lines of all pumps before any liquid or gas is allowed to pass through the pumps. This will trap any dirt in the system. Suction screens are normally placed in a flange or T-type strainer between the suction valve and the pump. The screens should be made of a fine screen retained by a heavy gauge, punched hole, cone-type strainer. Place these in the lines before pressure testing so that test liquid can be pumped through the lines.

- Do not carry out the hydrostatic test at temperatures that are low enough to cause the metal to become brittle. The minimum temperature for hydrotesting is typically around 16°C. If necessary warm the test water with steam.

- In some cases, vessels that have been shop tested cannot be hydrostatically tested in the field. (Typically, these are vessels that are designed and/or supported such that they cannot be safely filled with water.) They must be blinded from the process piping under test. For example, overhead piping in vacuum systems is often not designed for liquid-full conditions, and should be pressure tested in alternative ways.

- Particular attention should be paid to shell and tube heat exchangers where a leak during normal operation could put water or the heating fluid into the process stream. The tube rolls must be pressure tested very carefully. Check the exchanger vendor name plate test pressure regulations in cases where differential design is used on the tube bundles.

Procedures

- Divide the plant into zones.
- Determine the test pressure for each zone. It is frequently set at 1.5 times the cold design pressure.
- Isolate with blinds zones that are to be tested at different pressures.
- Close all low point drains throughout the system.
- Open all high point vents throughout the system.
- Water fill the vessel and sections of the system to be tested.
- Vent air at the high points until the system is liquid-full. Avoid air pockets in the system.
- Close all vents.
- Raise the pressure with a test pump to the prescribed test pressure.
- Disconnect the test pump.
- Hold the test pressure for at least one hour on major vessels and equipment.
- Hold the test pressure for at least half an hour on piping.
- If the pressure falls, check for water leaks from flanges and connections to determine the location of the leak.

Completion of the Test

On completion of the test, do not drain liquid too fast, otherwise a vacuum may form. Restrict the drainage rate to the capacity of the high point vents.

- Open the high point vents to the full open position.
- Drain the water.

PHASE 3 — COMMISSIONING

The distinction between precommissioning and commissioning is somewhat arbitrary. The convention used here is that, at the end of precommissioning the plant is mechanically complete and tested. Commissioning involves adding chemicals: first flush water, then process chemicals.

Representative steps in commissioning are shown below. Each plant will have its own steps, and the order in which they are carried out will vary from unit to unit.

1. System Flushing
2. Commissioning/Testing of Utilities
3. Pressure/Leak Testing
4. Nitrogen Purge
5. Refractory Dryout
6. Chemical Clean
7. Catalyst Loading/Activation

1. SYSTEM FLUSHING

The main purpose of the flushing step is to cleanse the equipment of construction debris, rust and scale. Generally, lines designed to handle liquids are flushed with water and drained. Lines designed to handle gas can be cleaned by blowing with air or nitrogen. If the lines and their supports are strong enough, they can be flushed with water.

The initial flush circulation through piping which contains control valves and heat exchangers will be established through their bypasses. The control valves and heat exchangers will be blocked in. They should be unblocked and flushed only when the flushing water is clean.

Before flushing is started, open the high-point vents on the suction side of the vessels to avoid pulling a vacuum when these vessels are drained.

Flushing is generally carried out by adding water to vessels and towers, and then using the process pumps or portable flushing pumps to pump the water around and out of the system. Both the service pump and the spare pump (when it exists) can be operated together to obtain the maximum circulation rate. It is possible to use a hydrocarbon for flushing. This permits higher velocities through the equipment without overloading the pumps. This generally results in a cleaner plant. In addition, the oil can be heated with steam. The resulting temperature increase will promote the flaking of mill scale. If an oil is used, the system will have to be steamed out following the flushing.

Debris will be flushed from drains and open flanges directly into the drainage system. Debris will also collect in the pump suction screens. These must be checked and cleaned regularly.

Dirty water may cause pluggage of internal items such as distributors and wire meshes. Therefore, it will be necessary to decide which vessels are to be flushed. In particular, heat exchangers should not be flushed until the circulating water is clean. To prevent dirty water from entering a vessel or an exchanger, break open the flange at the nozzle and wedge open the nozzle. Install a deflector plate or blind flange on the vessel side. This will cause the material to fall on to the ground. This is particularly important when flushing into vessel nozzles containing distributors or flushing at column trays.

Precautions

Observe the following precautions before starting the flushing process.

- Remove all orifice plates
- Block in all control valves - open their bypasses
- Blow all instrument air lines with clean, dry air. Then close off or disconnect
- Blind all relief valves
- Turn off all steam tracing
- Maintain accurate records of all flushing operations

General Procedures

1. Use as many lines as possible during the flushing operation to facilitate complete system clean-out.

2. Where possible, lines should always be flushed away from a vessel rather than into a vessel. (This is particularly important if the flushing process could put solid materials into internal

equipment such as distributors or tube sheets.) Any vessel that is being flushed into should be opened and manually cleaned following flushing.

3. Do not flush lines into exchangers. If flushing through an exchanger is necessary, clean it when flushing is complete. Where provided, open the circulating streams through exchangers and close exchanger bypass valves only when the circulating stream is reasonably clean.
4. Always flush through a bypass to an open end before flushing through equipment.

5. Always flush from highest elevation to lowest elevation.

6. Do not flush upwards if possible

7. When establishing flushing sequences, take care that debris from a dirty circuit is not flushed into a previously clean system.

8. Vessels and systems must be vented when filling or emptying. In particular, while draining, care must be taken to avoid a vacuum conditions.

9. All lines not flushed by vessel drainage or pump circulation must be flushed independently.

10. Flush through all drains and vents to check for blockage.

11. Achieve the highest flow rate possible.

12. Flush lines in the following sequences where possible:

    a) Main Header
    b) Each lateral header
    c) Each branch line

13. Continue flushing each line or piece of equipment until the discharge water is clear.

14. Use a unit piping flow diagram as a cleaning check-off record to ensure that all lines are flushed, including lines to and from tankage.

15. All towers and drums should be manually cleaned before flushing.

16. Flush out low-point drains on piping and vessels.

17. Where a flow rate is throttled to hold a level, periodically increase the flow rate as allowed, without overloading the driver.

18. Flush around control valves using their bypasses.

19. Lines not included in the pump circulation systems should be individually flushed with water, air or steam.

20. Rundown lines to other units or to storage should be cleared as close to their destination as possible.

Pump Operation

Starting the Pump

As debris collects in the temporary strainers, the flow of water may be restricted. If this happens, throttle the discharge valve of the pump, otherwise internal slippage within the pump could generate excessive heat in the pump casing.

- Before starting the pump, crack open the discharge valve of a centrifugal pump and fully open the discharge valve of a positive displacement pump.

- Start pump and bring it up to speed immediately, observing the discharge pressure. If pressure does not build up immediately, stop pump and find the cause. Refill the pump case and restart when previous difficulty is corrected.

- Always bring a centrifugal pump up to speed as quickly as possible. This will develop maximum discharge pressure as quickly as possible which will flush and lubricate the surfaces between the casing stationary wearing-rings and the wearing-rings of the rotating element.

- When the discharge pressure of a centrifugal pump has increased satisfactorily after starting, gradually open the discharge valve to obtain the desired flow rate.

- In the event of unusual noise, vibration, overheating, or other abnormal conditions, shut the pump down immediately. Correct the cause before resuming operation. Continue checking for abnormal conditions as these may occur after prolonged operation.

- Check shaft sealing. Mechanical seals should show no leakage. Conventional packed stuffing boxes must always be permitted to leak slightly to provide some lubrication and prevent overheating. Stuffing box gland nuts are generally only finger tight. A leaking mechanical seal must be replaced. Occasionally, a mechanical seal will have some leakage, due to slight misalignment on start-up, which will stop if the pump is started and stopped a few times

- Operate the pump and direct flows through all suction and discharge piping circuits.

- Inspect and clean suction strainers.

- Re-check and realign, if required, after any disturbance of piping, such as required for suction screen inspection.

- When one pump loses suction, shut it down and clear its screen.

<u>Shutting Down the Pump</u>

- For centrifugal pumps, close the discharge valve first. This maintains discharge pressure while the pump rolls to a stop. This also prevents reverse rotation after shutdown, in the event that the discharge check valve should leak.

- For positive displacement pumps, the discharge valves must not be closed until the pump is shut down and rolling stops, to avoid excessive discharge pressure.

Control Valves

Initially, control valves are blocked in, and their bypasses are open.

- When the system is clean, unbolt each control valve and roll it away from its flanges.
- Flush the control valve from both directions.
- Place the control valves in service.

Heat Exchangers

- Do not flush dirty water through exchangers. The debris may block the tubes.
- Where possible, establish circulation through heat exchangers with the exchangers blocked in and the bypasses open.
- If an exchanger does not have a bypass line, install a deflection plate at the heat exchanger inlets, as discussed under General Procedures.
- Place the exchanger into the circulation loop only when there is no longer any danger of debris being washed into the exchanger heads.

Completion

Flushing is complete when the pump screens stay clear for twelve hours.

- When the flushing process has been finished, check that all vessel vents are open.

- Dump the water.
- Close all drain and vent valves.
- Pressure with air to the specified pressure
- Use the air pressure to blow residual water out of low points in lines, exchangers, pumps and other equipment.
- Insert flow meter orifices in the clean lines once all flushing is complete.
- Open the flushing headers and each equipment connection.
- Blow with air to dry and clean the system thoroughly.
- Carefully check that all temporary breaks are reconnected, control valves replaced, blinds removed and pumps aligned (with the correct suction screens.)
- Replace all control valves and close their bypasses
- Close heat exchanger bypasses

2. COMMISSIONING/CHECKING OF UTILITIES

It is likely that many parts of the utility systems will have been commissioned before this phase of the operation. However, it is important to make sure that they are all fully in service before the start-up progresses any further. Utilities can include specialist services, such as amine loops in gas treatment plants.

If there is any reason to believe that a utility header is not clean, it should be flushed before commissioning.

Some suggestions for the steam systems are provided below

- Watch for thermal and mechanical shock, over-pressuring of casings and the introduction of condensate into other systems.

- Open all drains, vents and steam trap bypasses.
- Introduce steam into the system very carefully by slowly opening a block valve until steam is heard to just pass through the valve.
- Withdraw condensate through drains.
- Warm up the system slowly in order to avoid water hammer and shock.
- Check the traps regularly to ensure that they are operating. If they are not, condensate and water may accumulate in the steam piping.
- Drain any liquid from low points of the plant and instrument air headers, the fuel/natural gas headers.
- Keep drains and vents open until dry steam issues from each point.
- Throttle the steam valves slowly until the lines are at the desired pressure.
- Place all steam traps and condensate lines in service.
- Block in steam valves at vessels, snuffing steam, atomizing steam and turbine steam.

3. PRESSURE/LEAK TESTING

Toward the end of the commissioning phase, it is necessary to carry out a pressure test to ensure that there are no leaks at the flanges that have been opened during the flush-out process. This test is not as rigorous as the hydrotest.

It is usually done with plant air or bottled nitrogen at quite low pressures. Each flange and other potential leak point is checked with soap solution. The presence of bubbles indicates that there is a leak. This test should be done during daylight hours and when it is not raining.

Always carry out leak tests on equipment that was opened at any time before hydrocarbons are admitted to it.

4. NITROGEN PURGE

Once the unit is tested, air is purged from the system with an inert gas. Nitrogen is usually used for this, although fuel gas or some other hydrocarbon may be chosen. The plant will set a target for oxygen concentration of say 0.5%. In order to reach this value, it is often necessary to pressure up and then depressure three times. Nitrogen pressure does not usually need to go over 15 psig because this is not a pressure test (unless nitrogen is being used in place of air.)

5. REFRACTORY DRYOUT

The refractory lining of the vessels must be cured through a gradual high temperature dryout procedure. It is usually started with one pilot burner operating, then more burners are added as the temperatures rise. The dryout must be completed before the outdoor temperature falls below freezing. Details will usually be supplied by the Heater/Furnace vendor.

6. CHEMICAL CLEANING

Many plants will passivate the inside of the plant piping by pumping through a special acidic solution. This is almost invariably done by specialist companies, who will have their own procedures. The plant personnel are responsible only for showing the vendor where to make the connections to the process.

7. CATALYST LOADING/ACTIVATION

Commissioning includes loading catalyst into reactors, mole sieves and other process units. Often this has to be done under an inert gas atmosphere. Therefore full safety precautions have to be taken.

Following the catalyst loading, it may be necessary to activate the catalyst with the appropriate chemicals.

PHASES 4 AND 5 — START-UP AND LINE-OUT

Once the plant is commissioned, the formal start-up begins. The distinction made here between commissioning and start-up is that start-up involves the introduction of fresh feed and the production of final product. Line-out follows this: production rates are achieved, the product is on-spec and the plant is shown to meet the appropriate guarantees.

Phases 4 and 5 are part of Volume I — The Operating Procedures.

THE START-UP TEAM

The start-up team will usually consist of four groups.

1. The operating company which has bought the plant and which will be responsible for future operations.

2. The engineering company that was responsible for the detailed design and construction. Sometimes, these responsibilities are divided between different companies.

3. The licensor(s) of the process technology.

4. Equipment vendors.

THE OPERATING COMPANY

The operating company will usually provide the following personnel.

1. *Operations Superintendent*
   The Operations Superintendent has control of all operations. In many cases, he will be appointed the Operations Manager, once the start-up is complete.

2. *Maintenance Superintendent*
   The Maintenance Superintendent is responsible for keeping the unit operating once the mechanical completion step is finished, and the Prestartup Safety Review has been carried out.

3. *Shift Foremen/Operators*
   The shift foremen are responsible for immediate operations. The operators are the principal users of the manual (see Chapter 1.) They report to their respective shift foremen.

4. *Process Engineer(s)/Instrument/Electrical Engineer(s)/Mechanical/Piping Specialists*
   These specialists report as needed during the start-up.

THE ENGINEERING COMPANY

The engineering company will usually provide the following personnel. They correspond to their equivalents in the operating company.

1. *Chief Operator/Start-Up Supervisor*
   This person, who is not always provided, is the counter-part of the Operations Superintendent. He works as part of the start-up team, providing technical knowledge and general plant experience. He may also be responsible for the operating manuals and operator training.

2. *Process Engineer(s)*
   The process engineers report to the Chief Operator. They assist in the mechanical checkouts and in precommissioning. They may also help with training. Typically, they will go on shift during the actual start-up and will also be expected to handle process-related trouble-shooting issues. One of these process engineers could be a good choice to write the "As-Built" Operating Manual.

3. *Instrument/Electrical Engineer(s)/Mechanical/Piping Specialists*
   They also report to the Chief Operator and have duties similar to those of the process engineers.

4. *Construction Skeleton Crew*
   Often, the engineering company will provide a small crew of personnel representing some of the key crafts that were involved in the construction of the project.

LICENSOR(S)/EQUIPMENT VENDORS

The licensor will provide experts in the technology. Their job is to provide detailed guidance regarding the process. They are also a representative of their own company and can make sure that the plant is operated in accordance with the license.

Equipment vendors are brought in to assist with the start-up of their machinery or instrumentation and to provide specialist maintenance help as needed.

CHAPTER 6
WRITING AND PUBLISHING THE PROCEDURES

INTRODUCTION

The first five chapters of this book have described the principles behind operating procedures, and have provided some suggestions as to how they can be organized and formatted. This chapter discusses the writing and publishing steps. The importance of this phase of the project cannot be over-emphasized; if the procedures are not well-written, or if they are not attractively presented, the operators will not use them if they can avoid doing so. After all, most of the other materials that they read, such as magazines, books and newspapers, are usually professionally written and prepared — operators can reasonably expect their operating manuals to meet similar standards.

The appearance of the manual will also have a strong influence on the way in which outsiders and auditors view the quality of the product. This may be unfortunate, but outsiders will have trouble understanding the technical content of the manual because they do not know the technology of the process. Therefore they will place heavy emphasis on appearance.

Another reason for stressing the importance of the appearance of the procedures is that many of the people involved with writing procedures are technically oriented. In particular, engineers and scientists are trained to "let the facts speak for themselves." They are not encouraged to stress appearance and style; instead, they concentrate on originality, the development of analytical methods and integrity of data. Such an approach is inappropriate for writing operating procedures.

The advent of sophisticated desk-top publishing software, color printers and fast personal computers means that good-looking materials can be printed at a low cost. Therefore, there is no excuse for producing low quality material.

However, the incorrect use of these tools can create a different set of problems. Unless the users have had training in the graphic arts, they may turn out product that looks fancy, but that is difficult to read and use. This commonly happens with the over-use of color, or the indiscriminate mixing of different fonts. Consequently, it is safer for most writers of procedures to adopt a conservative approach to the publishing step, and to assign the creative, artistic side to someone who is trained in this topic, and who knows what they are doing.

The Reader

Before starting to write the procedures it is essential to determine who the reader will be, and to determine his reading skills, and the manner in which he is likely to use the operating procedures.

Method of Reading

It was pointed out in Chapter 1 that an Operating Manual is structured more like a data base than a book. This means that a person does not "read" the manual in a sequential manner; instead he jumps to those sections of the manual that are relevant to the tasks currently being performed, and then extracts the relevant information. Consequently, when writing the procedures, it must be recognized that the reader will often be reading chunks of the material on a stand-alone basis; it cannot be assumed that he has read all the related introductory and background information. Each module of the procedures must make sense on its own terms: it must be internally consistent and self-contained.

Reading Grade Level

The writers should know the approximate reading skills of the operators, usually in terms of grade-level equivalent. A target level of about eighth grade (roughly equivalent to the newspaper "USA Today") is sometimes used. This does not mean that the operators have a reading level of eighth grade; it is likely that they can read at a higher level. But writing the procedures at that level means that there is no need for the reader to concentrate on interpreting the syntax and grammar.

Various techniques for analyzing the readability of written material are available. For example, the "Fog Index" by Gunning [15] is calculated for a hundred word passage from the following equation.

$$G = 0.4 * (W + HW)$$

where G is the educational grade level of the material
W is the average number of words per sentence, and
HW is the percentage of hard words (all words with three or more syllables, with some exceptions).

There are other, similar techniques that can be used. In large part, they boil down to using simple words where possible ("eschew obfuscation.")

Modern word processors can scan a piece of text and automatically provide a value for the Fog Index or its equivalent, along with a grammatical analysis of the writing. Such packages will often make suggestions for improvements, although their suggestions tend to become rather repetitive and boring after a while. They can also be rather patronizing in tone, and sometimes they are simply wrong. Therefore, they should be used with judgment.

Finally, it must be noted that many operators have a high level of education, many of them possessing college degrees and/or teacher's certificates. It should not be assumed, therefore, that their reading level is inferior to that of the plant's support personnel; indeed, many operators will be substantially more literate than the engineers and managers.

LANGUAGES OTHER THAN ENGLISH

Throughout this book, it has been assumed that the procedures are to written for operators whose first language is English, and who can read English at a reasonable level, say eighth grade. In fact, many plants are run by operators who do not speak or read English fluently.

This topic is alluded to by OSHA in the Process Safety Management regulations (see Chapter 8.) They make it clear that it is management's responsibility to make sure that the procedures and the training program are provided in the language that the operators speak.

WRITING THE PROCEDURES

This book does not provide instruction on how to write English; such a goal would be presumptuous. It is best to read the classic by Strunk and White [16]. Nor is this a technical writing book. Many quality books on technical writing are available, for example that by Lannon [17] and the well-known *Chicago Manual of Style* [18] and Webster's Dictionary (for American English.) Nevertheless, some ideas as to the style of writing appropriate for use with operating manuals are provided here.

MINIMALIST WRITING

The computer software industry has developed a concept known as "minimalist writing" for use in their manuals and other documentation. The goal of minimalist writing is to present information in a manner appropriate to the way in which adults actually learn. There are various elements to minimalist writing, not all of which are useable, or even safe, were they to be applied to the process industries. Nevertheless, two of the ideas, the minimization of "soft" materials and the provision of a good index, are very applicable.

Minimize "Soft" Materials

The instructions in many operating manuals are cocooned in a cloud of "soft" materials. These may include introductions, summaries, and statements of the company's commitment to Total Quality. None of these materials really help the operator do his work better; instead they pad the manual with material that gets in the way and increases the time it takes to find useful information. In response to this, the minimalist approach eliminates as much as possible of these soft materials. In the case of operating procedures proper (Volume I) the manual should tell the operator how to operate his unit — no more, no less. All other materials must be banished to the other volumes of the operating manual, or should be purged from the manual altogether.

The inclusion of Title Blocks and Authorization Blocks in the operating modules increases the amount of soft material in the manual and so contravenes minimalist writing principles. This has to be counter-balanced against the benefits that these blocks provide.

Concise Writing

Minimalist writing is not quite the same as concise writing, although the two certainly have a good deal in common. Indeed, some writers make their procedures too concise with the result that they become difficult to read because the text is too cryptic and the words are crammed together. This gives the writing what is known as a "cobble-stone" effect, i.e. it bumps along from word to word and phrase to phrase with no connecting materials to smooth the flow. Shortness is not an inherent virtue in writing; there is nothing wrong with using sentences that flow smoothly and that use correctly-placed connective phrases and clauses. The goal is communication, not brevity.

This is one area in which the services of a technical writer can be very beneficial: they are trained to make written material flow smoothly and naturally.

Writing Styles

People who are good writers of normal business correspondence such as reports, letters and reports sometimes have trouble writing procedures because many of their normal rules and style guides do not apply. For example, it is generally considered poor style to repeat a word too often. But there is nothing wrong with this when writing procedures. Similarly, it is often considered poor style to use simple words such as "good" , "big" and "have" on the grounds that they are insufficiently precise. Writers are encouraged to come up with more descriptive words. This is not necessary when writing procedures.

Management Reporting

A management report usually consists of answers to the following three questions.

1. What are the problems that are being addressed?
2. What are the possible recommendations to solve those problems?
3. And what is the cost (financial, personnel and time) to implement the recommendations?

Many engineers, particularly those who serve as either external or internal consultants, are adept at this kind of writing. Unfortunately, it is not relevant to the writing of operating procedures.

Technical Report Writing

Technical writing is similar in style to management reporting. The author states a clearly-defined thesis, and then develops information that serves to support or negate the thesis. The essence of this kind of writing is a tightly structured organization and a rigorous logic. This approach is also inappropriate for procedures-writing because a manual is not intended to demonstrate the truth of a certain point of view or scientific theory.

Discursive Writing

Discursive writing flows without any explicit organization, such as the use of numbered paragraphs or headings and sub-headings. Generally, there is no place for this type of writing in operating procedures, although Volume II — The Operating Information can be written discursively.

EDITING

When editing a document, it is sensible to use standard proofreader's marks — a complete list of which is provided in the *Chicago Manual of Style* and in Webster's Dictionary. These marks will be understood by professional editors and printers.

PROTOTYPING

Although the top-down approach to developing a manual that was described in Chapter 2 is a good way of managing and controlling a procedures-writing project, it sometimes makes sense to develop one section of the manual in detail as a pilot project for the following reasons:

1. It provides management with a better understanding as to how much time and effort will be needed for the complete project.

2. It gives everyone, particularly the operators, a chance to work with, and to critique the format. It is much easier to comment on a worked example than to try and create something new. If this step goes well, everyone is much more likely to buy into the overall project.

3. It highlights any problems that are likely to arise, such as missing information or difficulties with the information-gathering process.

The publication of the prototype is the first time that the future users will actually see the product, so it is important that both the contents and the appearance are first-class. It should not be treated as "just a draft."

A fundamental requirement for successful prototyping is that the persons who prepare the first draft need to have thick skins and must not take criticisms (and there will be many) too personally.

PLAN TO THROW ONE AWAY

In his excellent book on the development of computer software *The Mythical Man Month*, Brooks [19] entitles one of his chapters: "Plan to throw one away; you will, anyhow." The idea behind this insight is that, no matter how well planned a software project may be, the changes that are made during its design and development will be so significant as to make modifications of the original structure almost impossible. Instead, the software writers will do much better to discard what they have done and to start again, with all the changes in specification incorporated into the new design. Although the idea of throwing away the first product sounds draconian and wasteful, it is, in fact, far more productive than trying to fix the existing structure, and it causes fewer project management problems than might be expected.

This insight can be applied to the writing of operating procedures. Many procedures-writing projects reach a point at which so many changes are required, and there is so much re-design of the modules and their linkage to be done, that it is easier to start again. This does not mean that all the previous work is wasted; indeed, the opposite is the case. All of the information that has been gathered previously will be used, and will be connected into the system in a useful and coherent manner. Usually the re-design focuses on the linkage between the modules, rather than on the modules themselves.

Writing Guidelines

This section provides a few suggestions or guidelines on effective writing for operating procedures.

Positive/Negative

Instructions are more likely to be understood and followed if they are written in a positive manner. For example, it is better to write

"Keep the flow rate below 2000 lb/h"

rather than

"Do not let the flow rate go over 2000 lb/h"

Active Voice

Generally procedures should be written in the active voice. For example: "The engineering department installed the drain valve in 1988" is better than: "The drain valve was installed by the engineering department in 1988." The first phrase, which is in the active voice, is more direct and it is shorter.

Passive writing of the type:

"The pump is started"

should be avoided because it is not clear whether this is intended to be an instruction or a statement. The phrase should either be

"Start the pump", or

"The pump will start"

Use of the word "should" is a particular problem. For example, the phrase "the pump should be started" could mean:

"Start the pump" or

"The pump is likely to be already running" or

"The pump might not be running" or

"Someone else will have started the pump"

"If you haven't started the pump yet, you're in trouble"

Another example is:

"The valve should be opened before measuring the temperature"

This could mean:

- That the valve can be expected to be in an open position — but it might not be — and it may or may not matter
- That the operator is expected to open the valve
- That someone else should have opened the valve

- That, regardless of the valve status, the temperature must be measured
- That the temperature should be measured only if the valve is in an open position.

This phrase is ambiguous in English, and could become hopelessly corrupted if it has to be translated into another language. Therefore, the instructions should be written in short clear sentences, starting with a verb in the imperative tense, and each instruction must be on a separate line. So the above example becomes two simple, sequential instructions.

1. Open the valve
2. Measure the temperature

Use of the Word "You"

Generally, the word "You" should be avoided because its use is not minimalist. It is better to write "Close the valve" than "You must close the valve."

Arabic Numbers

Numerical information should generally use Arabic numbers, i.e. '1'; not 'i' or 'I'. Use of Roman Numerals (I, II, III, IV, V, VI, VII, VIII, IX, X) can be confusing for people not used to this system. They are best avoided in operating manuals (although they can be used effectively in books and reports.)

Imperative Tense

Instructions should start with a verb in the imperative active tense, i.e. in the form of a direct order. The following sentence is correct:

"Open the valve"

The following would normally be incorrect.

"You must open the valve"

"Having opened the valve"

Avoid Superfluous Instructions

The instructions:

"Next, open the valve"

"Then, open the valve"

"Open the valve carefully"

could, in most cases, be replaced by

"Open the valve"

List Instructions Singly

Each instruction should be given its own sentence on its own line. So the phrase:

"After opening the valve, the pump should be started and the temperature measured"

should read:

- Open the valve
- Start the pump
- Measure the temperature

Implied Instructions

A common problem with manuals is that instructions are implied, rather than stated right out. For example, the phrase "Make sure that the pump is turned off" can be interpreted in one of two ways. Either it means "Turn off the pump" or it means "Check that the pump is turned off."

Bulleted Lists

Bulleted lists are an excellent way of organizing instructions, particularly if a modular table concept is not being used. Modern word processors offer many fine options in this area. For example:

- Instruction 1
- Instruction 2
- Instruction 3

The style of bullet can be changed:

⇒ Instruction 1
⇒ Instruction 2
⇒ Instruction 3

Conditional Instructions

The Department of Energy Writer's Guide [20] provides some rules for the use of conditional statements. It suggests that the use of the following words is acceptable.

IF, THEN, WHEN, AND, OR, NOT

The following words should be avoided.

FOR EXAMPLE, EXCEPT, BUT, ONLY IF

The Guide also suggests that conditional words should be written in underlined uppercase letters, as shown above.

The conditional phrase, and the action that follows it should be on separate lines. For example:

WHEN the temperature reaches 120°C

<u>THEN</u> line up the steam trap.

Tortuous or confusing logic should be avoided. For example, the following instruction can be interpreted in one of two different ways.

<u>WHEN</u> the temperature reaches 120°C <u>AND</u> the flow is 66,000 lb/hr <u>OR</u> the pressure is 100 psig

<u>THEN</u> line up the steam trap.

This would be better written as follows:

<u>WHEN</u> the temperature reaches 120°C <u>AND</u> the flow is 66,000 lb/hr,

<u>THEN</u> line up the steam trap.

<u>OR</u>

<u>WHEN</u> the temperature reaches 120°C <u>AND</u> the pressure is 100 psig,

<u>THEN</u> line up the steam trap.

It is even better would be to simplify the operation itself so that there is no longer any need for the operators to have follow this complex logic.

Much of the above guidance on the use of conditional words is written for the nuclear power industry. Generally, the process industries do not use such a tightly structured and formalized approach. Nevertheless, the clear logic paths that result from use of methods such as those described above does provide a good framework for the operating procedures.

Emphasis Techniques

The most commonly used methods for emphasizing text are CAPITALIZATION, **bold**, <u>underline</u>, *italic*, emphasis boxes, color, font changes, and white space. They are generally used to highlight critical information. Once a method of emphasizing the written material has been selected, it should be used consistently throughout the manual. No more than 10% of the material should be emphasized [21].

Figure 23 provides illustrations of some of these emphasis techniques.

Figure 23

Emphasis Techniques

CAPITALIZATION
Generally, capitalization should be avoided except to emphasize single words or very short phrases. Strings of upper case letters are hard to read [22] and should be avoided.

Bold
The use of bold printing should normally be avoided because it might not be sufficiently distinct, particularly if the toner in the laser printer is low.

Underline
The choice of underline as an emphasis technique is generally good. There is no doubt about the emphasis, it is clear to see, yet it is not intrusive.

Italic
The use of italics is effective but, like bolding, it might be overlooked if the printer does not make the font distinctive enough.

Emphasis Boxes
Emphasis boxes are used for Warnings, Cautions and Notes.

Color
This topic is discussed in a later section of this chapter.

Spelling

There is no excuse for mis-spellings, nor for misspellings. All modern word processing packages incorporate comprehensive spell-checkers: they should be used. Technical and plant-specific words can be added to the spell-checker.

Most word processors also contain a thesaurus. This will suggest alternative words that have similar meanings to the one just entered. Not only will this sometimes provide a better choice of word to express an idea, it will also liven up the writing by adding variety to the vocabulary.

The correct spelling of potentially troublesome words that are frequently used in operating manuals is listed below. In all cases the Webster dictionary was used.

- *Start-up* (hyphenated, even though many people, including OSHA, make it one word)
- *Shutdown* (one word)
- *Troubleshooting* (one word)
- *Precommissioning* (one word)
- *Stand-by* (hyphenated)

Lists

Volume II (The Operating Information) will often contain lists for items such as equipment or alarm settings. These lists should generally be written in an indented form rather than as a continuous string of text. For example:

The following alarms will enunciate during a reactor high temperature trip:

FAL-124	Low flow of feed to the reactor,
TAH-127	High intermediate bed temperature
PAH-121	High interstage cooler pressure.

is much better than the sentence:

The following alarms will enunciate during a reactor high temperature trip: FAL-124 Low flow of feed to the reactor, TAH-127 High intermediate bed temperature, and PAH-121 High interstage cooler pressure.

Writing Fluency

Writer's block is a common and serious problem with people who are charged with writing procedures. It is important to make the written words flow fluently and easily, yet this is difficult for many people.

One way of improving writing fluency is to organize as much of the material as possible at the start of the project. This is one of the advantages of a modular design; much of the writer's work consists of little more than filling in the blanks.

Use of Humor

Humor should be avoided for the following reasons.

1. Humor relies on surprise to make its effect. Operating procedures are intended to be used frequently, therefore the surprise effect will be lost and the humor will become tedious and distracting.

2. Humor does not tell an operator how to carry out a task; therefore, it contravenes the principles of minimalist writing.

3. Humor can have unfortunate, unintended side-effects because what is funny to one person may be tasteless or offensive to another. People of a different ethnic and national background from the writer of the procedures may interpret the humor in ways that were never intended but which, nevertheless, cause offense. Even simple technical words can have unanticipated meanings.

4. Humor can be perceived as being patronizing.

5. Humor can trivialize the importance of the instructions.

Jargon and Slang

Although the use of day-to-day slang and jargon should be avoided, there is no reason that plant-specific jargon need be excluded if it helps to describe an operation more clearly and if it is not offensive. After all, the procedures are written for the operators. This means that they should be written in the language that they actually use during their working day. It is important, however, to be sure that they all use the same slang terms.

Repetition of Messages

The writers of an operating manual have to decide how much repetition of a message is appropriate. For example, if a text-graphics approach is used, there is some built-in redundancy. Generally, it is desirable to have at least some repetition for the following reasons.

- A second message will have some differences from the first one. The difference will provide the reader with a better understanding of the instruction. This is particularly true if one message is text and the other is graphics.

- If one message contains errors, it will show up as a discrepancy when compared with the other.

- Repetition will help reinforce critical statements.

DANGERS, WARNINGS, CAUTIONS AND NOTES

There needs to be a consistent and systematic method of highlighting problems and dangers, particularly for those situations where a single error could cause some type of accident. The following terminology — shown in Figure 24 — can be used. It is derived from military standards.

Figure 24

Dangers, Warnings, Cautions and Notes

DANGER — Incorrect action could lead to serious injury, or death. It could also cause a major environmental problem or massive equipment damage. Should the **DANGER** message be needed, the hazard should almost certainly be eliminated. It is unacceptable to have a situation where an operating error could lead to a fatality or serious injury, and where operating procedures are intended to provide the only safeguard.

WARNING — Incorrect action is likely to lead to injury and/or serious equipment damage or production loss. Whenever a WARNING message is required, consideration should be given to removing the hazard, rather than relying on procedures and training to provide defense.

CAUTION — Incorrect action may lead to a minor injury or minor economic losses.

NOTE — A note indicates that care is required or that important information is being supplied. Failure to carry out these instructions should not lead to an accident or significant process upset.

All messages to the operator must precede the instruction with which they are associated. There is no value in telling someone about the hazards associated with an action after they have taken that action.

Warnings, Cautions and Notes should provide information only. They should not contain actual operating instructions. This is something that is often overlooked.

One way of designing Warnings, Cautions and Notes is illustrated in Figure 25.

Figure 25

Examples of Warnings, Cautions and Notes

WARNING !!!

THIS IS AN EXAMPLE OF A WARNING

!! CAUTION !!

THIS IS AN EXAMPLE OF A CAUTION

NOTE

This is an example of a Note

As shown, these messages can be illustrated with graphic images. Images such as this are common in safety signs. However, graphics are not always helpful. It may require more time to interpret them than to read a few words. Also graphics can have the same sort of problems associated with the use of humor — they can be perceived as being patronizing and/or they can disrupt communication.

PUBLISHING THE PROCEDURES

Once the procedures have been written, they are printed and distributed. This publishing step is very important, and is one that does not usually receive sufficiently high priority. However, the increased availability of high-power, low cost desk-top publishing systems makes this an area where great improvements can be made with a relatively modest investment. It is important, however, to use modern printing techniques with care and conservatism. Printing is a specialized subject, in which most procedures-writers are not trained. Generally, it is best to be cautious and conservative.

LAYOUT/FORMAT

Layout refers to the location of text (copy) and graphics on the page. Format refers to physical issues such as typeface, margins and column widths. In practice, the two terms tend to be used interchangeably by non-professionals.

Line Length

For discursive material such as that provided in Volume II (Process Information), the line length should be about 40 characters long. If they are longer than this, the line spacing should be increased.

White Space

It is important to provide plenty of white space in the manual. People do not read text word-by-word, they read groups of words together. The correct use of white space will help the reader create the right groupings of words, and so facilitate the reading. White space also serves as a formatting device; if a section of text is isolated from the text surrounding it by a large amount of white space, the reader knows that it is being highlighted. Even when the written material is not being emphasized, text density should not exceed 40% [23].

Fonts

A font is set of letters, numbers, punctuation marks, and symbols that share a unified design. The choice of font will depend on what pleases the users (and possibly on company standards.) However, two *caveats* should be kept in mind.

Serif Font. A serif font is one where the letters have small extensions (such as those in this book.) These should generally be chosen because they make the text seem to flow more naturally. *Sans serif* (without serif) is often a better choice for titles and headings. This sentence is printed sans serif with the Arial font.

For many writers of manuals, the font of choice is Times New Roman (which is serif.)

Point Size. The point size of the font is a measure of how large the letters are (the larger the number, the larger the text.) Most procedures will be written in 12 point or larger. If a smaller font is used, the manual may be difficult to read in a work environment, particularly if it has to be read from a distance. Headings, titles and check boxes may be in larger font, say 18 point.

The body of the text in this book is 11 point Times New Roman.

Mixed Fonts. Technically, it is quite easy to put two or more different fonts on the same page. However, this should be done with trepidation. Contrast is best obtained by mixing emphasis and size within the same font.

Paragraph Formatting

Modern word processors all give the option of justified printing. This means that the words on a line are spaced so as to completely fill the line. Because some letters, such as 'A' take up more space than others, such as 'j', the amount of space allotted to the letters is automatically adjusted so as to create a pleasing effect. Nevertheless, justified text can still be a strain to read, particularly when there is a small number of long words on a line. Therefore, many writers choose "ragged right margins" in order to eliminate these potential problems. This book is written with ragged right margins.

Another choice is whether to start each paragraph with an indent. If the indent is not used, as with this book, some space is saved. Otherwise, it is really a matter of taste.

Heading Styles

For those parts of the manual, such as Volume II, which are discursive, it is useful to have a standardized heading style. Modern word processors can automatically create the correct formatting for each heading. An example of heading styles is shown in Figure 26.

Figure 26

Example of Heading Styles

LEVEL 1 — 16 POINT, UPPER CASE, BOLD, CENTERED
LEVEL 2 — 14 POINT, SMALL CAPS, BOLD, CENTERED
LEVEL 3 — 12 POINT, SMALL CAPS, SINGLE UNDERLINE, LEFT JUSTIFIED
Level 4 — 12 Point, Bold, Italic, Left Justified

Page Numbering

It is important to have a page numbering scheme so that, if any pages are removed, their loss is immediately apparent. For those sections of the manual which are modular in structure, there is no need for an overall page numbering scheme. Indeed, as discussed in Chapter 2, such a scheme would actually be an impediment because the systems is designed so that modules can be added, and broken apart.

Single-Sided versus Double-Sided Printing

It is necessary to decide if the manual is to be printed single-sided or double-sided, i.e. whether the reverse side of each piece of paper is to be left blank. The advantage of single-sided printing is that pages can be changed very easily because there is only one module of information on them. Also, photo-copying machines tend to be mechanically unreliable when they are printing double-sided copies.

If they fail to copy both sides correctly, large quantities of paper will have to be trashed or recycled.

Double-sided printing saves paper. It lends itself well to the two-page module approach that has been described throughout this book.

INDEXING

A good index is an important part of any manual, yet it is rarely provided. There are two types of index: the Table of Contents at the beginning of the manual, and the Topic Index at the end. The Table of Contents shows the organization of the manual; it does not provide much detail but it does give the reader an idea as to what the major topics are, and where they are to be found. The index identifies the location of keywords.

Index cross-referencing can also be provided within the operating modules. This is one of the purposes of the "Response/Discussion" column discussed in Chapter 2.

Each piece of information in the manual should be indexed in as many different ways as possible so that it can be accessed by different approaches. The first way is to use different phrases that refer to the same item. Consider the following ways of referring to P-100.

	Module #
Starting Pump P-100	200.2.21
How to Start Pump P-100	100.1.12
Pumps - Starting, P-100	500.6.01
P-100 - How to Start	500.6.03

This item can also be linked with other sections of the manual using the "See," "See Also" approach, shown below.

	Module #
How to Start Pump P-100	200.2.21
See Also: Pumps, general principles	Volume II, page 38
Overamping pumps	Volume II, page 32
Cavitation of pumps	Volume II, page 32

Regardless of how the index is constructed, it is important not to wind up with circular references where two index items refer to one another.

One indexing problem to watch for concerns graphics. The graphics page may contain pictures of equipment that are not referenced directly in that procedure but that "just happen to be there." For example, if the graphic is a picture of Pump P-100, Vessel V-120 may show up in the background. Although V-120 has nothing to do with the operation referred to in the procedures, it must be indexed so that, should it change in any way, that picture can be pulled out of the procedures and updated. Further information on methods for creating an index can be found in the *Chicago Manual of Style.*

GLOSSARY

A glossary is used to define technical terms and abbreviations. It will not always be required. The best location for it is in Volume II. The glossary should not overlook simple terms because it may be used by an outsider, such as an auditor.

THE USE OF COLOR

Most operating procedures are published using black ink on white paper. However, the advent of low-cost color printers and color photo-copiers is changing this. The sensible use of color highlights important information in the procedures, and makes them look more attractive (this is important because it will encourage the operators to use them.) It also enhances the quality of the information, making it easier to follow a sequence of steps. Color can also be used to identify original documents (assuming that the company does not use color copiers). If an original has say a red title block, and is then copied, the copy will not show the color, and so is identified as not being an original. This is a technique that some quality standards, such as ISO 9000, can use to identify any illicit photocopies.

Color should be used sparingly, and should be used only to contrast different sections of the page. Color should not be used (in operating procedures) merely to make the document look interesting or attractive.

Some of the warnings concerning the use of sophisticated publishing techniques by people who have not been trained in their use are particularly relevant with respect to the use of color. Too much color, or the use of colors that clash with one another will detract, not enhance, the appearance of the procedures.

It is important to choose colors that contrast with one another tastefully. This is usually illustrated with the color wheel. The major color points, moving clockwise, are Violet, Blue, Green, Yellow, Orange and Red. Complementary colors (those from opposite sides of the color wheel) should not be placed next to one another. Doing so can cause a perception of shimmer or vibration. Selecting colors that are near one another on the color wheel creates a sense of harmony. Colors that have three other colors between them provide a very strong sense of contrast.

Color photographs are generally a better choice for graphic illustrations than black-and-white, even when the final, published copy is to be monochrome, because a process plant tends to offer a very sharp contrast between the dark outline of the equipment and the light background (often the sky.) This has the effect of creating a profile, with the equipment being in deep shade and all detail being lost. Color pictures provide for a smoother transition; the contrasts are not so extreme.

THE BINDER

The first impression of the procedures is the binder, therefore it is worth devoting considerable attention to its design. This will often include the company logo and some appropriate graphics.

It is suggested that the Emergency Procedures be put in their own volume (Volume III)) and that the binder for that volume should have a distinctive, bright color (usually red.) This will allow the operators to locate it quickly, at a time when they are under considerable stress. It will also provide the information that they need without it being cluttered up with other, non-emergency operating information. Included in the emergency operating procedures are plans for activities such as evacuation, major fires, communication with other refineries and agencies, and public relations.

Where funds permit, it is suggested that the binder should be silk-screened, with customized graphics on both the cover and spine.

LOCATION OF THE PROCEDURES

Generally the operating procedures will be located at a convenient point in the control room. However, there are alternative ways of distributing them.

LAMINATED CARDS

Laminated cards are very convenient to carry around. They can be useful where an operator may need to refer to a checklist. They can also be very useful as a summary of emergency operating information.

In addition to having laminated cards that can be carried around, larger versions can be printed up and attached to the equipment item that they refer to. Being laminated, the cards should be able to withstand outside weather conditions.

Hip-Pocket Guides

Hip-pocket guides contain a summary of key operating information. They are printed on very robust or laminated paper. The control of these guides for updates can be a problem.

On-Line Documentation

Operating procedures are usually written with software such as word processors, graphics programs, CAD programs and desk-top publishers. The next logical step is to consider distributing the manual electronically, using either stand-alone PCs in the control room, or adding the procedures to the plant's DCS (Distributed Control System.) The paper manual then becomes a backup to the electronic system.

On-line documentation is not the same as simply putting the written procedures on a disk, and then displaying them electronically [24]. The product itself must change to reflect the peculiarities of the new medium. Given that on-line documentation is much more than an electronic display of written documents, it is useful to compare the two in more detail.

Sequential versus Random Access

Electronic documentation should provide the reader with multiple access points and multiple views of the information contained in the procedures. Indeed, in a well-designed system, the structure and the layout of the procedures can, to a certain extent, be defined by the user. The organization of the information can be somewhat dynamic. This is not the case with paper manuals.

Maintenance

- There is no need to worry about the manuals becoming lost, damaged or defaced if they are stored electronically. New copies can be issued very easily.

- Updates and changes can be transmitted quickly and easily. The draft version of the changes can be transmitted to the control rooms for comment using the same system.

- It is cheaper to store information on disks than on paper.

Ease of Use

Although on-line documentation and electronic distribution of the operating procedures has many advantages, there are also some drawbacks. One of the biggest problems is that the reader often has trouble knowing where he is in the overall context of the manual. With a written book, this is usually quite easy. A glance at the number of pages on either side of the one that is being read quickly provides a sense of location. This is not the case with a computer screen. You cannot "curl up with a good computer screen."

Another problem with electronic documentation is that it is not so easy to simply browse through the manual or to flip pages looking for the item in question.

Integration with the DCS

If the procedures are to be distributed electronically, it would make sense to incorporate them into the Distributed Control System (DCS.) At the time of writing, the typical DCS does not have sufficient capacity to do this. However, it is reasonable to assume that this will change in the next few years.

COMPUTER HARDWARE

This section discusses the selection of computer equipment for writing and publishing procedures. Computers are already so powerful, and their cost is so modest, that hardware should present no limitations to what is needed. In general, however, it is necessary to have top-of-the-line systems. As computers become more powerful, so the software is developed to take advantage of that power; and the writers of operating manuals are most certainly "power writers."

COMPUTERS

The operating procedures will usually be written on a computer using word processing software. It is unlikely that there will be any special computer requirements; any modern PC or mini-computer network should be capable of handling the project. However, an important requirement is that the computer's operating system be one for which powerful and friendly word-processing software is available. If the procedures are to be distributed electronically, the choice of computer may be more restricted.

There is a trend towards using Graphical User Interfaces (GUIs) for all kinds of computer work, including word processing. This allows the writer to see the words as they will look when they are printed. This feature is useful because it allows the writers to visualize the layout of their work as they are writing. The drawback is that this type of interface can be slow. Moving a mouse to a menu, then moving down the menu, then clicking on the item of choice is much slower than pressing one or two keys (referred to as a command-line interface). Consequently, computers that use GUIs can slow down the power user, unless command keys are also available.

PRINTERS

Any modern laser printers should be adequate for the task of printing the pages. A resolution of at least 300 dots per inch (dpi) is necessary to obtain good-looking copy. 600 dpi is better if the printer is to be used for printing graphics or CAD drawings. For those pages that are to be printed in color, it may be necessary to use an ink-jet printer because color laser printers are quite expensive.

It is important not to overlook the type of paper used. The impression made by any printed document depends in part on the quality of the paper on which it is printed. When a person picks up a document that is printed on good quality paper and says that "it looks nice," he is probably reacting to the actual feel of the document. It is possible that what he should really be saying is that "it feels nice." Therefore, it is very important to give due attention to the quality and color of the paper selected. Also, the paper must be strong enough to hold up in the environment in which the manuals are used. If the manuals are taken outside, for example, consideration should be given to what happens to the paper if it becomes soaked with rain, or splashed with chemicals.

Scanners and Digital Cameras

Optical scanners are used to electronically record information that is only available on paper. The information can be either graphics or text. Once it has been scanned into the computer, the image can be rotated, partitioned, and zoomed. Optical scanner software often includes a feature that attempts to identify words that it could not properly read. So, if it picked up the letters 'op~rator', it would suggest that the word is 'operator.' The person using the scanner can either accept or reject the computer's suggested spelling improvements. Scanners are often useful for reference material that has already been written, say for an engineering study, and that needs to be inserted into Volume II, the Information Manual.

Unfortunately, the optical character recognition software that scanners use can make errors, some of which are egregious. It is extremely important to have a person proof-read the final text before it is released.

Digital cameras take pictures that are recorded directly in an electronic form. Their use avoids the need to create half-tones of photographs which then have to be scanned into the manual.

Software

Publishing Software

Operating procedures are usually written with a word-processor on a Personal Computer (PC) network. The decision as to which software package to use is usually affected more by the overall company environment than by the needs of the operating procedures project. Realistically, most modern word processors are so powerful and are capable of integrating files from so many other packages (including graphics, CAD and spreadsheets) that any package should be suitable. Nevertheless, when choosing the word processing software, the following factors should be kept in mind.

Top-of-the-Line

The persons writing the procedures should always have the very latest and most powerful software packages. They are power users and will use all the latest features.

Tables

The modular concept makes heavy use of the tables capabilities of the software. Therefore, it is important that the tables can be quickly created, edited and modified. This is important to the writers, because they are power users and who need to be able to write and edit quickly.

Numbered and Bulleted Lists

One very useful feature of modern word processing packages is the ability to create automatically numbered lists. Given that many operating procedures are fundamentally lists of instructions, this feature can be very handy, particularly when editing lists. For example, a procedure may contain the following set of instructions:

1. Measure the temperature of the liquid in T-1200.
2. If the temperature is greater than 160°C, start P-1200.
3. Pump liquid into T-1201 until level in T-1201 is 50%.
4. Stop P-1200.

It is then decided to add another instruction to go between steps 2 and 3, as shown.

1. Measure the temperature of the liquid in T-1200.
2. If the temperature is greater than 160°C, start P-1200.
3. If P-1200 cavitates, stop it immediately.
4. Pump liquid into T-1201 until level in T-1201 is 50%.
4. Stop P-1200.

The person who made this modification forgot to change the old step 4 to step 5. If the automatic list numbering feature had been used, this change would have been made automatically.

Color

The software should offer a wide range of color options, both on-screen and to the printer. Color should be printable in both the foreground (the text) and the background.

Speed

Procedures-writers must be able to move around the document quickly. The software should offer single or double-key commands for all the commonly used features, such as changing formats and editing tables. All unnecessary on-screen clutter, such as "Power Bars" and "User Chummy" help should be capable of being turned off.

Thesaurus

A good thesaurus provides a range of optional words. This is useful because it can add precision to the instructions and it can make them more lively by reducing the amount of repetition.

DOS versus Windows

For writers who are working with the PC family of computers, there is often a choice between using DOS-based and Windows-based software. As with other aspects of computers and software, the decision can be quite emotional at times. However, leaving aside feelings, it is probably fair to say that DOS-based systems are good when a lot of material is to be typed quickly. The software is usually much faster because relatively little computer power is needed to display the document. Windows-based packages are more appropriate when it is important to see what the final document is going to look like.

It is possible to have the best of both worlds. The bulk of the writing and documentation can be entered using a DOS system. Then the Windows-based system can import the DOS file that has been created and convert it so that it can be edited under Windows.

Project Management Software

The topic of project management is discussed in the next chapter. The software that is usually needed is a Critical Path Scheduler and a spreadsheet for tracking progress.

Graphics and CAD

The layout of the modules described in Chapter 2 was to put text and graphics on separate pages that face one another. One of the reasons for this is that it allows the writers to edit the text and the graphics separately. However, modern word processing packages and desk-top publishing programs make this less necessary, the graphics software files can be imported directly or the graphic can be created directly within the word processing software.

A graphics package is used to draw sketches in an electronic format. CAD (Computer Aided Design) is a special form of graphics where the lines that are entered are always given exact dimensions. For this reason, CAD programs are used for engineering graphics. As mentioned in Chapter 1, many CAD programs possess an "attributes" capability. This means that components on the drawing, such as pumps and valves, have attributes assigned to them concerning their physical properties. But the attributes may contain more abstract information, such as information derived from a HAZOP and the operating procedures for running that item. The use of attributes permits a CAD program to become the foundation for a complete technical information management system.

CHAPTER 7
MANAGING A PROCEDURES WRITING PROJECT

INTRODUCTION

The previous chapters of this book have described some of the technical aspects of operating procedures. This chapter discusses the management of a procedures-writing project. Fundamental to the ideas expressed in this chapter is the realization that a project for writing procedures is indeed a full-scale project in just the same way as say installing a large piece of equipment is a project. Therefore, there needs to be a budget, a schedule and a project manager. The project also requires a substantial commitment of resources, particularly of experienced plant staff. Although this may sound obvious, some companies do not treat the writing of procedures in this manner. Instead, they attempt to fit the activity in with their other work. The writers try to get their manuals prepared during their "spare time," and the operators are asked to review the draft products during quiet times on their shift. This is both unreasonable and ineffective.

This chapter is divided into two major sections:

1. Development of the Writer's Guide
2. Development of the Project Manual

THE WRITER'S GUIDE

The Writer's Guide provides instructions on how the manual is to be written. Essentially, this book is an extended Writer's Guide. Table 11 provides a representative Table of Contents for the Writer's Guide.

Table 11

Representative Table of Contents for The Writer's Guide

I. **INTRODUCTION**

II. **MANUAL ADMINISTRATION**
- A. MANUAL ORGANIZATION
  - 1. Procedures
  - 2. Information
  - 3. Emergencies and Trouble-Shooting
- B. VOLUME I — OPERATING PROCEDURES
  - 1. Modular Structure
  - 2. Design of a Module
  - 3. Linkage of Modules
- C. VOLUME II — OPERATING INFORMATION
- D. VOLUME III — EMERGENCY OPERATING AND SHUTDOWN PROCEDURES
- E. VOLUME IV — TROUBLESHOOTING GUIDE
- F. VOLUME V — INITIAL START-UP
- G. DISTRIBUTION
- H. MANUAL REVISION
- I. ANNUAL CERTIFICATION

III. **COMPUTER HARDWARE/SOFTWARE**
- A. COMPUTER HARDWARE
  - 1. Computers
  - 2. Printers
  - 3. Plotters
  - 4. Scanners/Digital Imaging
- B. COMPUTER SOFTWARE
  - 1. Word Processing
  - 2. Graphics
  - 3. CAD

Table 11 (continued)

IV. WRITING GUIDELINES

- A. WRITING RULES
  - 1. Syntax
  - 2. Grammar
  - 3. Writing Style
  - 4. Font
  - 5. Point Size
- B. MODULE LAYOUT
- C. INDEXING SCHEME
- D. PAPER SIZE
- E. FORMATTING
  - 1. Design of Modules
  - 2. Linkage of Modules
- F. HEADING FORMAT
  - 1. Level 1
  - 2. Level 2
  - 3. Level 3
  - 4. Level 4

V. PRINTED MATERIALS

- A. BINDERS
- B. PHOTOGRAPHS
- C. DRAWINGS
- D. P & IDs

VI. TRAINING

The Project Manual

This chapter is written on the assumption that the plant for which the procedures are being prepared already exists and that a considerable amount of actual operating experience is available. The project will be managed somewhat differently if the plant is new and has yet to be commissioned. In particular, there will not be an experienced operator who is intimately familiar with that particular unit; consequently the role of the resident engineer, who validates the procedures at the time of plant startup, becomes particularly important.

The Project Teams

It is suggested that operating procedures should be written by teams, rather than by a single individual who performs all of the tasks in the project. The team approach will ensure that specialist skills are utilized most effectively; also interaction between the team members will improve the quality of the final product, and will maximize employee participation. Therefore, the first step in managing a procedures-writing project is to decide on the function and composition of the teams. The approach described in this chapter is to use two teams.

The first of these, *The Resource Team*, is responsible for developing the technical information that is to be included in the manual. The second team, *The Publishing Team*, is responsible for editing and publishing the final product. If a company decides to use a contractor to help with the project, the contractor will most likely fill the role of the Publishing Team and/or the project management. Both of these teams report to a single project manager, who is in overall control of the project. The organization chart for this management system is shown in Figure 27.

Figure 27

Representative Organization Chart

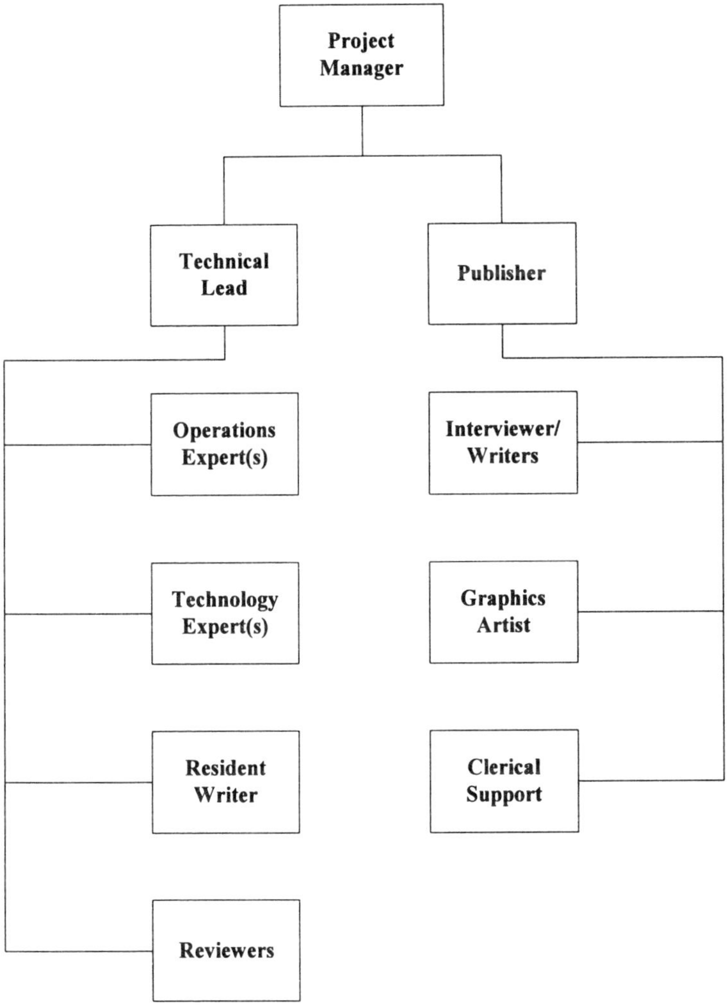

Both teams should work on site, if possible. When there are questions to answer or technical problems to solve, they can often be resolved simply by walking on to the unit and inspecting the appropriate equipment or by talking to the operators. However, many process plants are located in relatively remote areas, whereas publishing facilities are usually in large urban areas. In these cases, it is inevitable that some members of the Publishing Team will be working off-site.

The following sections provide detail as to the composition of the teams and the skills required of each individual. What is more important than any specific skill, however, is the ability to work with others. Procedures-writers must be able to get along with all types of persons. If a person cannot work with a team, he should not be put on the project, regardless of his technical or writing skills.

Project Manager

The project manager is in charge of the overall project. He is responsible for all facets of the work, including technical quality, financial control and scheduling. He does not necessarily have to be an expert on procedures or on the unit being described.

In practice, the overall chart will probably be somewhat smaller, with one person filling two or more of the roles. But, they should always understand the distinct nature of these roles.

The Resource Team

The resource team provides technical and operating skills for the project. The following personnel should be assigned to the team (in many cases, one person will cover more than one role, nevertheless the distinct nature of the respective roles should be understood.)

The Technical Lead

The resource team technical lead is responsible for organizing the personnel necessary to provide technical and operating information. He is also responsible for the contents of the manual and for ensuring that the material in it is accurate, complete, relevant and up-to-date. He has authority over the publishing team's work if he feels that the technical integrity of the procedures is not of a sufficiently high standard. In many cases, the roles of Project Manager and Technical Lead will be held by the same person, even though their responsibilities are distinct. The technical lead may also be one of the operations experts.

Operations Expert(s)

The key member of the resource team (assuming that procedures are being written for a plant that is already up and running) is the operations or subject matter expert. This is the person who knows the unit in great detail. This person is often a supervisor or senior operator. A retiree often makes a good choice; he will not have scheduling conflicts, and can be brought into the project on an as-needed basis. Ideally, there should be more than one operations expert on the team because the interactions between the two experts will generate useful insights. It is assumed that the Operations Experts will not be responsible for the actual writing of the procedures because most operators and supervisors are not trained in writing, nor do they enjoy the writing process.

If the plant for which the procedures are being written is new, or has been substantially altered from its previous mode of operation, operations experts will be supplied from other units that use similar technology.

Technology Expert(s)

Technology experts are those people who understand the basic theory and design of the process. They do not necessarily have a detailed knowledge of how the plant operates, particularly during startup and shutdown, but they can provide insights as to why a certain operation works in the way that it does. They are particularly useful for establishing operating limits. The technology expert will often be the process engineer assigned to the unit, but assistance may be provided by other departments such as maintenance, instrumentation and environmental.

The technology experts are not usually skilled at writing operating procedures because, like the operations experts, they have not been trained in how to organize and write a manual. Also, procedures written by technical people often over-stress the informational and technical parts of the manual, but fail to provide sufficient information as to how to actually run the plant.

The technology experts are sometimes called *Subject Matter Experts* or *SMEs*.

Resident Writer (for new plants)

No matter how good a set of operating procedures for a brand-new plant might be, there will inevitably be errors and missing items that only become apparent once they are used to actually operate the plant. For this reason, it is suggested that an engineer or some other technically qualified person be on hand at the time that the procedures are used for the first time. His job is to carefully evaluate the accuracy and the usability of the procedures. The resident writer will follow through each task with the operators and will read and evaluate what is written in the instruction and log books. A resident writer is particularly important when the procedures are being written for a new plant because the team will not have had the advantage of working with an operations expert.

Essentially, he will be creating a Volume I from Volume V.

Reviewers

Once the first draft of the procedures has been prepared, qualified reviewers are needed to evaluate it for technical content and ease of use. There are usually four types of reviewer:

1. The person assigned to check the procedures as they are being written
2. The operations personnel who will be using the final product
3. Management (because they have to sign the authorization boxes
4. Outside auditors

The first three of these types of reviewer do their work as the procedures are being written. Their aim is to assist the writers, particularly with technical accuracy and completeness. The fourth type of reviewer is the auditor. The formal audit process is conducted by an outside party using some objective standard, such as a regulation or a company policy. Although auditors will help improve the quality of the procedures by making constructive comments, they are often viewed in the short term as adversaries.

The way in which the review process is managed is very important. There are two potential problems:

1. The reviews do not get done in a reasonable time, or

2. The review process goes around and around without someone calling a halt to it.

In the consulting business there is a well-known saying that "Draft Reports Don't Exist." (Another proverb that expresses the same idea is "You never get a second chance to make a first impression.") This insight stems from the fact that consultants, whose final product is usually a report, are often under considerable pressure to issue a first draft in the early phases of the project. The clients say that will ignore any superficial problems or omissions and that they will discard the draft once they have read it. This rarely happens. The first draft is often photo-copied and widely distributed around the client company. It can make an indelible impression on the readers such that, no matter how many subsequent corrected and improved updates are issued, they fixate on the first draft. Any mistakes or omissions are pounced upon and remembered for ever. The same principle applies to operating procedures.

There will be considerable pressure to issue an early first draft of the manual for two reasons. First, people will want to have a look at what has been developed so that they can check it for technical accuracy, even though the manual, being a draft copy, it is certain to contain errors.

The second reason for issuing an early draft is that management is usually investing considerable financial and human resources into the project and, human nature being what it is, they wish to know what is going on and to make their own desires known.

In some ways, it is right that drafts should be issued quickly so that people can become involved in the project and feel a sense of participation and ownership. This is one justification for prototyping, as described in the previous chapter. Therefore, one of the project manager's on-going problems is to resolve the tension between issuing drafts that are so preliminary as to cause more harm than good, and issuing them so late that people are not fully involved in the project. There is no simple answer to this dilemma; but the project manager does need to be sensitive to it.

Another decision concerning the release of the first draft is whether or not to issue it piecemeal. It is suggested here that, as far as possible, this should be avoided. It is important that the readers and reviewers see the procedures as an integrated whole. If individual modules are released, the reader will not have a good overall impression of where they fit in; thus, he is likely to become confused and unsure of the overall picture.

The best way of reviewing the procedures is in a team atmosphere. Ideally, all of the unit's operators will be present. (This is not usually realistic unless the plant is undergoing a complete turnaround.) Nevertheless, the more operators that can be present during the review the better.

The structure of a team review of this type is very much like a HAZOP (see Chapter 8.) There is a team leader, who understands process plants very well but who is not necessarily an expert in the process being reviewed and is not necessarily the lead on the procedures-writing project. However, he has a lot of operating common sense, and can analyze the procedures, and make sure that they follow generally accepted practices. There is a team secretary or scribe, who takes notes. The secretary is technically knowledgeable and not only understands what is being said but participates in the discussion. The remainder of the team consists of operators and management representatives. The team analyzes the procedures module by module (where the modules correspond roughly to the nodes in a HAZOP.) They also review the relationship between the modules, i.e. the sequence in which the plant is started and stopped.

The worst type of review is one in which the procedures writer drops a copy of the manual on the operator's desk and ask him to "take a look at it when you have a moment." This virtually guarantees that it will not be reviewed in sufficient depth. It also flies in the face of the principle of employee participation which, as has already been stressed, consists of much more than simply communicating with the employees.

In practice, one of the biggest snags in getting the operating procedures finished is the management review step. One or more managers has to sign the modules before they are released. If this is to be done responsibly they will need to take a considerable amount of time to satisfy themselves that the work was done well and that the procedures are accurate and usable. But the people who have to do this are usually very busy with many of their more urgent, day-today activities. Consequently, they often become a bottleneck.

THE PUBLISHING TEAM

The publishing team is responsible for the design, layout and preparation of the manuals. Their job is to make the procedures attractive and easy to use. This team will usually consist of the specialists listed below. As with the Resource Team, one person will sometimes cover more than one assignment.

Publisher

The Publisher is responsible for ensuring that the procedures are typed, the illustrations prepared and the final product printed in an attractive format. He will work with the Technical Lead and other team members on the design and organization of the manual. The publisher reports to the project manager.

Interviewers/Writers

The interviewer is the person who obtains the information from the process and operations experts. He will generally have a good working knowledge of process plants in general but does not need to be an expert on the process in question. This person must also be trained in technical writing. Having obtained the information from the operator, the interviewer then writes it down in modular form.

The interviewer asks the experts how they operate their plant. The questioning can be quite open-ended and relatively unstructured. Once the knowledge has been captured, the interviewer will then organize the information into a structure of linked modules. He will then print out an early draft of the procedures and ask the process and operations experts to review it. This process can be repeated a few times until the experts are satisfied that what is written down is an expression of their intent. This writer will also collect information from other sources, such as existing procedures and engineering manuals. If the operators actually write the first draft of the procedures, it is the writer's job to organize and edit them.

The advantage of this approach is that the process experts are not required to actually *write* the procedures (unless they want to.) The process of writing is not one that they usually enjoy or that they are good at. However, the experts usually enjoy talking about their work, and they have little trouble in reviewing the written work of others.

Graphics Artist

The previous chapters of this book have stressed the importance of graphics. The development and layout of practical and tasteful graphics is a specialized subject, and not one in which most people working in a process plant have been trained. Therefore, it is suggested that a specialized graphics artist be included in the team, at least on a part-time basis.

Clerical Support

Preparing operating procedures usually requires a substantial amount of typing and clerical work. Typists can be brought in on a temporary basis for this activity.

Project Management

It has been stressed throughout this book that a project for writing procedures should be treated like any other project, such as the installation of a large item of capital equipment. This being the case, the management of a procedures-writing project is essentially the same as any other type of project management.

Goals

The project team first has to decide how they know when the procedures are completed. This is not easy because procedures can never be truly "complete," there are always changes and improvements that can be made to any manual. Therefore, when developing the Writer's Guide and the Project Manual, it is important to provide as much guidance as possible as to what constitutes "completion." This is particularly important if the operations are being upgraded in conjunction with the writing of the procedures. The project team has to have clear guidance as to whether they are describing the operating "as it is" or "as it will be."

The Mythical Man-Month

In previous chapters, reference has been made to the book *The Mythical Man Month* by Frederick Brooks. He states that there are five reasons that software programming projects fall badly behind schedule. They are:

1. Methods of estimating are poorly developed
2. Effort is confused with progress
3. A lack of firmness in telling senior management that the project cannot be speeded up
4. Poor monitoring techniques
5. Adding more people to reduce the slippage

Brooks further points out simply writing a program is only a part of the whole project. If it is to be useful, the program has to be robust enough to be used by a wide range of anonymous users, many of whom will mis-use the product. Therefore, extensive testing is required.

A second level of complexity arises when the program is fit into a larger system. It must be designed to fit with many other programs, to accept their input and to create results that they can use. The estimate is that each of these levels of complexity increases the total effort by a factor of three. Therefore, writing a program that is robust and that fits into a larger system takes at least nine times the effort of simply writing the initial piece of software.

All of these problems are pertinent to the writing of operating procedures.

A particularly common fallacy is the assumption that adding more staff to a project will speed it up. This may not be true because most tasks demand a finite amount of time, regardless of how many people are helping. Furthermore, as more people are added to the project, they spend more and more time communicating and training one another rather than doing direct work. Hence, the effect of more people has the effect shown in Figure 28. Initially, more people speed up the job. Then it makes no difference, and finally it actually increases the total time.

Figure 28

Effect of Adding More People

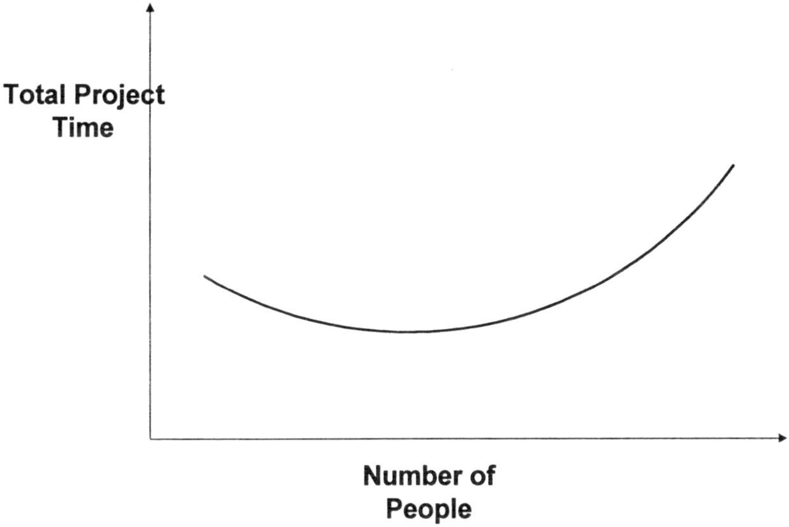

In the case of operating procedures, the most serious constraint often becomes the number and/or availability of the best subject matter experts.

Tracking System

One method of tracking a project is illustrated in Table 12. The Table was created for a plant containing five operating units (Units 100-500), a Utilities Area and a Tank Farm. The Table is then divided into three major sections: Operating Phase, Operating Limits and Safety & Health. Once more, these correspond to the requirements of the standard. The major cost of any procedures project is peoples' time. Therefore, Table 12 shows effort in terms of work-days. Other costs, such as computer equipment and travel, can be added later.

Table 12

Project Tracking System

SUMMARY SECTION						
Modules	**Required**	**414**	**Complete**	**210**	**Fracti on**	**0.51**
Work-Days		**1221**		**560**		**0.46**
	Modules Required	*Modules Complete*	*Fraction Complete*	*Work-Days Required*	*Work-Days Used*	*Fraction Complete*
Operating Phase						
Initial Startup						
Unit 100	29	21	0.72	87	63	0.72
Unit 200	11	2	0.18	33	6	0.18
Unit 300	63	23	0.37	189	69	0.37
Unit 400	41	41	1	123	123	1
Unit 500	18	18	0.61	54	33	0.61
Utilities	28	28	0.07	84	6	0.07
Tank Farm	3	3	1	9	9	1
Normal Operation						
Unit 100	20	12	0.6	40	24	0.6
etc.						

Table 12 (continued)

Temporary Operations						
Unit 100		1	1	5	2	0.4
etc.						
Emergency Operations						
Unit 100	3	3	1	6	6	1
etc.						
Operating Limits						
Unit 100				30	2	.07
etc.						
Safety/ Health						
Chemicals				2	0	0
Precautions				1	0	0
Contact Measures				2	0	0
Safety Procedures				18	160	0.89
Quality Control				3	0	0
Special Hazards				1	0	0

For each section, the number of modules and the number of work-days needed for each element is provided. The work-days are divided into the effort needed to write the modules and the effort needed to work out the linkage and structure. Initially, these numbers will be rough estimates but, as the project progresses, they will be made more concrete. The fifth column shows how many work-days have been expended so far on that element and the sixth column shows fractional completion. At the top of the spreadsheet is a summary section. It shows the total effort needed, the effort expended so far and the percentage completion. This is shown for both the number of modules and the work-days required.

So, for example, Table 12 shows that it is estimated that 29 modules will be required for the initial startup of Unit 100. To date, 21 (72%) of these modules have been completed. It is estimated that the modules will require a total of 87 work-days, and it shows that so far 63 work-days have been used. Overall, 210 modules have been written (out of a total of 414) and the project is 51% complete.

This spreadsheet is very simplified. For example, it does not show actual completion dates, nor does it break down the module development into its constituent parts (interviewing, writing, receiving, publishing.) Nevertheless, it provides management with a sound appreciation of progress to date, and how much effort is still required.

Filing and Disk Management

One advantage of the modular concept is that different people can work on different elements of the project at the same time. However, it is important to set up a system that ensures that only one person can be working on a particular module at any one time, otherwise there could be a considerable amount of confusion. If the procedures are being written on a mainframe computer system, control can be established by setting up a system of passwords that allow only authorized persons to access certain modules.

Control is more difficult to achieve if the procedures are being developed on a system of stand-alone PCs. An effective method of maintaining control is to have two sets of disks: *working* and *master*. Anyone writing procedures uses the working disks, of which there is only one set.

The key to this system is to make sure that there is only one set of working disks. Then, whoever is working on them, precludes anyone else from working on that same topic. So, for example, if one disk is labeled Unit 100 - Temporary Operations, only the person who has that disk can work on that topic. When they have finished their work, they will return the disk to the central file, thus making it available for anyone else who has material for that section. All other copies of modules in that area are unauthorized. If the changes that have been made are approved, they are transferred to the master disk.

The same concept applies to filing of written materials. There should be one official file for say Unit 100 - Temporary Operations. All authorized materials, such as photographs and interview notes, should be kept in that file. At the end of each working day, the Publishing Manager must ensure that all the working disks are backed up. Also all written materials should be backed up on a regular basis.

If procedures are being developed on a local area network, it is important to have control measures that prevent different writers from working on the same files and interfering with one another's work. A convenient way of doing this is to create sub-directories on the network: one for each volume or section of a volume. One person is given control of that subdirectory and the working volume that resides on it. If anyone else wishes to add material to that section, they e-mail it to the person responsible for the section. That person then adds it to the master document. Also, on the same sub-directory, will be the master volumes.

Development Of Volume I — The Operating Procedures

The section of the procedures that will usually take the most effort is Volume I, the Operating Procedures Manual. For this reason, additional detail on its development and management is provided in this section. It is suggested that Volume I can be developed in seven steps.

Step 1 — Design the Layout
Step 2 — Name and Link the Modules

Step 3 — Issue the First Draft
Step 4 — Write the Modules
Step 5 — Develop the Graphics
Step 6 — Review and Authorize
Step 7 — Publish

These steps are not in strict sequence with one another. For example, if a key person is only available for part period of time, it may make sense to write some of the detailed modules before the final linkage in another part of the unit has been resolved.

Step 1 — Design the Layout

The first step in the development of Volume I is to design the overall structure, i.e. to develop a Table of Contents. This is critically important because it provides the framework for the overall project. As has already been discussed, the manual should always be "complete" in the sense that the overall framework is in place, even if there is very little detail. Step 1 is the development of this framework.

Step 2 — Name and Link The Modules

The next step is to decide what modules have to be written and how they are to link with one another. This is an interactive process. For each section of Volume I (Start-up, Shutdown, etc.) a team should be assembled to work through the sequence of events as they take place. If, for example, start-up is being discussed, the team will describe what is to be done, who is to do it and when they are to do it. The team recorder will take notes and will try to divide the activities into module-size sections. He will also show how the modules link to one another sequentially. This process of naming modules and determining their linkage to one another is carried out many times until the team has developed a consensus.

One way of carrying out this process is to use the yellow "sticky" notes, used for inter-office correspondence. They stick lightly to whatever they are attached to, and they can be easily moved and attached somewhere else without damaging the original document. The team recorder will write down the name of each module on a sticky and place it on a large piece of paper attached to the wall of the meeting room. He will then draw arrows connecting one sticky to another to correspond to the order in which the actions are executed. As the discussion progresses, the stickies are moved around, and new stickies are added. Once this process has been completed, the final layout is written down.

Although a diagram such as this identifies the modules, it is likely that some of them will be too big and that they will need to be broken down. In order to keep track of the modules, and the children of the modules, it is important to follow the module numbering system.

Step 3 — Issue The First Draft

Once the modules have been identified, and their linkage developed, a first draft of Volume I should be issued for comment. What the reviewers will receive is a framework and a design; they will not receive much detailed information at this stage.

Step 4 — Write the Modules

The next step is to write the detailed modules. This is usually done by the interviewers and operations experts working together.

Step 5 — Develop the Graphics

Once the modules are written, the graphics (photographs, iconic flow diagrams, P & ID fragments) are prepared.

Step 6 — Review and Authorize

At this point, the manual can be issues to the formal reviewers and the management and supervisors can sign their names in the Authority Boxes.

Step 7 — Publish

This topic is discussed in detail in Chapter 6.

THE COST OF WRITING PROCEDURES

The largest single cost of writing procedures is the time of skilled people. The total time that these people need will depend on the following factors.

- The number of units for which procedures are being written,
- The size of each unit,
- The complexity of each unit,
- The quality of the existing procedures,
- The level of detail of the procedures,
- The scope of the project (for example, whether there is to be a Troubleshooting Guide),
- The published quality of the final product.

Once a company has developed experience with writing procedures, it should be possible to develop a "complexity index" that can be used to predict the cost of future projects. This is analogous to the Nelson index used in refineries for estimating the capital cost of different units. Process hazards analyses can also be costed using a variety of different factors, such as age of the plant, number of equipment items and the hazard level of the chemicals being processed.

Table 13 is a spreadsheet that shows what effort might typically be involved in the development of a single module. If a sheet such as this is prepared for each module, they can be linked to one another to make a living document that can be used to track the overall project cost.

Inspection of Table 13 shows that the cost of this example module is estimated to be in the region of $1,500. The labor rates used in this example are shown. Naturally, these rates for actual projects will vary widely depending on who is doing the work and whether internal or external resources are being used, and whether they include overhead and/or profit.

Table 13

Example of a Module Cost

	First Draft	Artwork	Review	Publish	Total
Hours					
Interviewer/Writer	4	2		1	7
Operations Expert	4		2		6
Clerical	2			2	4
Technical Expert			1		1
Management			1		1
Graphics Artist		2		2	4
Cost, $					
Interviewer/Writer	300	150		75	525
Operations Expert	300		150		450
Clerical	50			50	100
Technical Expert			100		100
Management			100		100
Graphics Artist		100		100	200
TOTAL					**1475**
Rate, $/h					
Interviewer/Writer	75				
Operations Expert	75				
Clerical	25				
Technical Expert	100				
Management	100				
Graphics Artist	50				

If outside contractors are to be used, they must be part of the team in a long-term sense. It does not make sense to simply hire a "technical writer" unless that person is meant to do no more than simple editing. The contractors need to be part of the team so that they understand the processes being written about, and so that they understand the company's requirements.

Table 14 provides a representative Table of Contents for the Project Manual.

Table 14

Representative Table of Contents for the Project Manual

A. SCHEDULES
B. RESOURCES
C. BUDGETS
D. QUALITY CONTROL
  1. The Review Cycle
  2. Operator Feedback
  3. Management of Change
  4. Field Checks
E. THE WRITING TEAM
  1. Outside Help
  2. Project Manager
  3. Resource Experts
  4. Publisher
  5. Writers
  6. Clerical Help
F. THE REVIEW TEAM
G. REPORTING STRUCTURE
H. DOCUMENT CONTROL

CHAPTER 8
PROCESS SAFETY MANAGEMENT AND OPERATING PROCEDURES

INTRODUCTION

Throughout the first seven chapters of this book, there have been many references to OSHA's Process Safety Management (PSM) standard, particularly as it applies to operating procedures. Regardless of whether a company is required to meet the specific requirements of OSHA's standard, the intent of the regulation is inherently sound. It makes sense for any company to follow its guidance.

Most companies will be writing operating procedures at the same time as they are carrying out other elements of PSM. Therefore it is important to understand how a PSM program works, and how an operating procedure-writing program fits in with it.

FEDERAL REGULATIONS

This section provides a brief overview of the way in which federal regulations are issued.

Once the United States Congress has passed a law, the appropriate government department and agency then develops a regulation to meet the intent of that law. Industry is then expected to comply with the terms of the regulation. In the case of 29 CFR 1910.119, the law which provides its authority is the Amendments to the Clean Air Act (CAAA). The regulation is administered and enforced by the Occupational and Safety Administration (OSHA), which is part of the Department of Labor.

Federal Regulations are listed in the "Code of Federal Regulations" (CFR.) The CFR is divided into numeric Titles; the pertinent Title in the case of Process Safety Management is **29**, which covers **Labor**. Title 29 is further divided into Parts, numbered from 1900 to 1910; part 1910 covers PSM. The parts are subdivided and identified with suffix numbers. The number 119 covers the "Process Safety Management" standard. Therefore, the regulation is identified as CFR **29** 1910.119. The sections of each Part are also grouped under letters. In this case the letter 'H' is the one that is used, signifying Hazardous Materials. These subpart letters are not often used; however, if they are, the full title of this regulation is CFR **29** Subpart H 1910.119

The OSHA PSM regulation is a federal regulation. Therefore it applies to all states in the union, unless a state has its own corresponding standard which is at least as stringent.

FUNDAMENTALS OF PSM

MANAGEMENT

Process safety management is concerned with the process of *managing* a plant so as to reduce the frequency and consequence of accidents. In particular, it is concerned with eliminating those accidents that occur as a result of unintended deviations from design and/or operating intent. Therefore, PSM is largely concerned with ensuring that no unauthorized changes are allowed to take place either inadvertently or intentionally.

Because PSM is a management standard, it is not fundamentally about safety as such. It is about making sure that the managers are in control of their plant. If management is in control, then safety will follow. Furthermore, good management systems will lead not only to improvements in safety, but also to other benefits, including increased production and improved environmental performance. In this context, a manager is taken to be anyone who has control of a process and who could therefore lose that control and thus let an accident take place. Therefore line operators and maintenance workers are managers.

PSM is not the responsibility of a single department, such as safety or operations. It involves all individuals (full-time and contract) and all departments.

PSM is one form of a Total Quality Management (TQM) program [25]. When integrated with other means of loss control, such as Reliability Centered Maintenance (RCM) [26] and Probabilistic Risk Assessment (PRA) [27], PSM can become an even more formidable tool in reducing losses of all kinds: safety, economic and environmental.

Performance Based/Non-Prescriptive

PSM is *performance based*, it is focused on results, not actions. This is why PSM regulations are generally *non-prescriptive;* they do not provide detailed, specific instructions on how to manage safety. The reason for this is that every process plant is unique. Consequently, no single safety standard or management system is suitable for all plants. Instead, management in each plant and each company has to work out how they are to implement PSM in their own facilities. This means that the regulation can only provide a framework and guidance; management and line personnel have to work out the details for themselves. For example, the regulation does not state that the procedures have to be developed in five volumes, as suggested in this book; it simply says that the operators have to have procedures whose content and quality is such that they are able to carry out their work safely. In one plant this may involve the use of five volumes, in another it may require just one volume. Management, with the full participation from the employees, has to decide.

Thinking Process

In Chapter 1 it was pointed out that this book provides guidance rather than rules for writing operating procedures. What is important is that the people writing the procedures think through the issues and develop a response that fits their situation. The same philosophy applies to all aspects of Process Safety Management.

Compliance

This view of PSM means that the only way to be in "compliance" with the regulation is to avoid accidents [28]; simply implementing a program is not sufficient. This also means that, should there be an accident, the plant is automatically in non-compliance, regardless of the quality of its PSM program. Indeed, given the above definition, a plant can never be fully in compliance because there is always a chance that an accident will occur, no matter how well PSM has been implemented.

For these reasons, it is probably best to avoid the word "compliance" altogether when implementing Process Safety Management. Probably the only time when the word is relevant is when a plant is formally inspected by OSHA or some other regulatory agency. The decisions made by the inspector at that time are effectively a *de facto* compliance standard.

Overall, it can be seen that there is really nothing new about the PSM. All companies follow its requirements already — if they did not they would be having many serious accidents. The regulation changes only two items:

1. It is now a legal requirement
2. The standard provides a well-thought out framework around which to build the program.

The fact that the concepts covered by the regulation are not new is important for another reason. As PSM is being implemented, many of the people involved will complain that the time it is taking is not allowing them to get on with their "real work." The response to this complaint is that PSM activities always *were* their real work; it should not have been a new activity. If this is not the case, then the problem is one of *old*, not *new* management practices. The high work load that a PSM program can generate occurs because there is so much catching-up to be done.

INTEGRATION OF THE ELEMENTS OF PSM

PSM is holistic: each of the elements interact with one another, and cannot be properly managed in isolation. For example, if it is decided to work on the Emergency Response Plan, the following sequence may occur. The element numbers refer to the OSHA standard.

1. The *Emergency Response Plan* (Element 12) requires a knowledge of what hazards have to be addressed. Consequently, a
2. *Process Hazards Analysis* (Element 3) is required. But, if this is to be conducted properly, much of the
3. *Process Safety Information* (Element 2), such as Piping and Instrument Diagrams (P & IDs) need to be in place. And, as the Emergency Response Plan is being prepared all members of staff have to be involved. This is
4. *Employee Participation* (Element 1). The Emergency Response Plan then has to be
5. *Audited* on a regular basis (Element 13.) Once the plan is complete, it has to be incorporated into the
6. *Operating Procedures* (Element 4), and its contents imparted to all affected individuals as a part of the
7. *Training Program* (Element 5).

This simple example involves interaction of at least seven of the elements of PSM.

The order in which the elements are tackled is important. It is probably best to start by making sure that the P & IDs are complete and accurate. They form the technical foundation of almost all the other elements, so they must be correct. Following completion of the P & IDs, the next element to address is Management of Change. If this is not operating at the start of the project, all the other elements (including the P & IDs) run the risk of being unreliable because of changes that effect them.

PSM MANAGEMENT

It is suggested that a PSM management book be put together. It would have fifteen sections, one for each element and one for a summary of the total effort. Each section of the book would have three principal sections.

- *Goals and Objectives.* This section would describe the companies goals for that element of the standard. As has been stressed for operating procedures, there can be no one single way of meeting the standard's requirements; all plants are different.

- *Compliance Matrix.* The standard for each element is listed line by line, and the response that the company has made is matched with the standard. Such a matrix can be developed for each element.

- *Discussion.* This section provides discursive material to show what is being done to meet the standard.

Many of the management techniques for writing operating procedures that have been discussed in the first seven chapters of this book can be used for all the elements.

ELEMENTS OF PSM AS DEFINED BY 29 CFR 1910.119

There are a number of PSM standards, in addition to that from OSHA. Some states, such as New Jersey, Delaware and California have PSM standards. There are also non-regulatory standards, such as the American Petroleum Institute's Recommended Practice 750, which covers this topic. All these standards are quite similar to one another. Therefore, a program which addresses one is likely to meet the others (although the legal aspects must be carefully double-checked.)

The following discussion is based on OSHA's standard, which contains fourteen elements. A brief description of each element is provided, along with a discussion as to how it integrates with the writing of operating procedures. The elements are:

1. Employee Participation
2. Process Safety Information
3. Process Hazards Analysis (PHA)
4. Operating Procedures
5. Training
6. Contractors
7. Pre-StartUp Safety Review
8. Mechanical Integrity

9. Hot Work
10. Management of Change
11. Incident Investigation
12. Emergency Planning and Response
13. Audits
14. Trade Secrets

EMPLOYEE PARTICIPATION

Employee participation is fundamental to the success of a PSM program, therefore it is appropriate that it is the first element of the regulation. Its aim is that all employees (whether they be hourly, salaried or contract workers) be full participants in the program's development and implementation. For example, many of the operators and maintenance workers will participate in HAZOPs, the development of P & IDs, and training and emergency drills.

Employee participation is much more than merely telling the employees what is going on. The standard talks about *participation*, not about *communication*. If this element of the standard is to work (and if it does not work the whole PSM program will fail) the employees must develop and run the PSM program. It belongs to them, not to management or a specialist PSM group. The role of the senior managers is to provide an environment in which the employees feel that they can indeed fully participate. This means that the managers have to remove organizational blocks that get in the way of the employees. A company which merely informs its employees about what it is doing about PSM has misunderstand the nature of this element. The employees *are* the program; which is why it is so important that operators be full participants in the development and review of operating procedures.

The skillful application of widespread employee participation is one of the most effective ways of reducing the costs associated with the implementation of PSM. This is because PSM work can be integrated with other tasks and will not require its own separate budget, nor will it require such extensive use of outside help.

PROCESS SAFETY INFORMATION

The purpose of this element is to make sure that all of the information that the operators need to do their job safely is available to them. The details of what information is required is spelled out in the regulation. It is always the responsibility of management to make sure that the information that is provided to the operators and other personnel is pertinent for *their* plant, regardless of what works for other plants, even if they are in the same company.

There is no requirement that all of the information be located in one place. As has been described in earlier chapters of this book, much of the process safety information will be located in Volume II of the Operating Procedures (assuming that the format suggested here is the one that is used.) It is particularly important that the operators have access to P & IDs and to variable value ranges.

PROCESS HAZARDS ANALYSIS (PHA)

The purpose of a process hazards analysis is to make sure that the hazards associated with a unit are recognized, understood and controlled. OSHA also makes it very clear that they expect action to be taken in a prompt manner once hazards are identified.

TYPES OF PHA

There are many types of PHA, including Hazard and Operability Studies (HAZOPs), Failure Modes and Effects Analysis (FMEA), the Qualitative Fault Tree, Checklists and the "What If?" technique. Each has its appropriate niche, depending on the nature of the plant and the type of analysis that is needed. A brief overview of one of the most widely used of these, HAZOPs, is provided here because the HAZOP technique can be extended to provide a means of checking the accuracy and completeness of operating procedures. It also provides a source of information for the Emergency Procedures and the Troubleshooting Guide.

At the heart of a HAZOP is the idea of "deviation keywords," as illustrated in Table 15 (an actual HAZOP may use more words than shown here.) Some of the cells, such as "Reverse Temperature," have no physical meaning, so they are blanked out. The HAZOP team considers each of the major variables (flow, pressure and temperature) at each section of the process ("nodes") and determines what could cause that variable to deviate high and deviate low. For each scenario, an estimate of the consequences is provided. Where appropriate recommendations are also generated. The purpose of a recommendation is to:

a) remove the hazard entirely, or
b) mitigate the consequences of the hazard, or
c) reduce the probability of the hazard actually occurring.

Table 15

HAZOP Table

	Flow	**Press ure**	**Temp**	**Loss of Contain ment**	**Level**	**Human Factors**	**Siting**
High (More/Too Much)							
Low (Less/Too Little)							
No		*(Vacu um)*					
Reverse							
Miscellane ous							

HAZOPs and Procedures

HAZOPs traditionally tend to concentrate on equipment issues such as the reliability of check valves, the design of instrument loops and the pressure rating of equipment. HAZOPs are not usually used to analyze operating procedures. This is unfortunate because, in a typical operating situation, it is estimated that about two thirds of upsets and accidents stem from human error, or at least the consequences of the accidents could have been much less if the human response had been more effective. Although HAZOPs do not usually discuss human error, there is no reason why the cannot be used, in a modified fashion, to do so. However, a non-conventional set of keywords is required. Some examples of these are:

- **Missing**. A step in the procedures is missing.
- **Skip**. The step is in the procedures, but its execution is omitted.
- **Wrong Order**. Procedures are carried out in the wrong sequence.
- **As Well As**. Some other action is performed at the same time as to the one in question, and the combination leads to some type of accident.
- **More/Less**. These words would cover actions carried out to the wrong degree, such as opening a valve too wide.
- **Other**. This covers human factors problems. For example, an operator may be confused between two gauges that look the same.

If a company needs to upgrade their procedures and also to perform HAZOPs on the same unit, it is very effective, both in terms of technical results and logistical efficiency, to conduct both projects in parallel. The work will be divided into the following phases:

Phase 1 — Prepare the first draft of the procedures. They should be at the level where they can be used, but they will not necessarily have been validated or issued, and they will not necessarily have been published in an attractive format.

Phase 2 — Conduct the HAZOP. The draft procedures are available to all the team members. Whenever a discussion concerning potential operating error arises, the team can refer to the procedures. If any changes are required, the team secretary can note them down.

Phase 3 — Review the updated procedures with a team of experienced operators. They will almost certainly generate some recommendations both for upgrading the procedures and for reducing hazards.

Operating Procedures

OSHA requires that the operators are provided with accurate, useful operating procedures. This is, of course, the subject of this book. The text of the operating procedures section of the regulation is quoted verbatim in the first column of Table 16. The second column of the same table provides discussion as to how that section of the regulation might be interpreted and addressed. References to Volumes I through IV are based on the structure described earlier in this chapter.

Table 16

The Operating Procedures Element

(1) The employer shall develop and implement written operating procedures that provide clear instructions for safely conducting activities involved in each covered process consistent with the process safety information and shall address at least the following elements.

(i) Steps for each operating phase:

(A) Initial startup;

(B) Normal operations;

(C) Temporary operations;

(D) Emergency shutdown including the conditions under which emergency shutdown is required, and the assignment of shutdown responsibility to qualified operators to ensure that emergency shutdown is executed in a safe and timely manner.

(E) Emergency Operations;

(F) Normal shutdown; and,

(G) Startup following a turnaround, or after an emergency shutdown.

(ii) Operating limits:

(A) Consequences of deviation; and

(B) Steps required to correct or avoid deviation.

(iii) Safety and health considerations:

Table 16 (continued)

(A) Properties of, and hazards presented by, the chemicals used in the process;

(B) Precautions necessary to prevent exposure, including engineering controls, administrative controls, and personal protective equipment;

(C) Control measures to be taken if physical contact or airborne exposure occurs;

(D) Quality control for raw materials and control of hazardous chemical inventory levels; and,

(E) Any special or unique hazards.

(iv) Safety systems and their functions.

(2) Operating procedures shall be readily accessible to employees who work in or maintain a process.

(3) The operating procedures shall be reviewed as often as necessary to assure that they reflect current operating practice, including changes that result from changes in process chemicals, technology, and equipment, and changes to facilities. The employer shall certify annually that these procedures are current and accurate.

(4) The employer shall develop and implement safe work practices to provide for the control of hazards during operations such as lockout/tagout; and confined space entry; opening process equipment or piping; and control over entrance into a facility by maintenance, contractor, laboratory, or other support personnel. These safe work practices shall apply to employees and contractor employees.

TRAINING

OSHA requires that the operators be properly trained in the processes that they are running so that they have full understanding of what they are doing. There is also a requirement that the training process be properly documented.

If the operating procedures are as thoroughly written as is suggested in this book, it is questionable if there is a need for a separate training manual on how to run the plant. If the operating procedures are supplemented with a training program that features on-site demonstrations, classroom sessions and the use of video tapes, there should be no need for further documentation.

There is an important distinction between training and education. A person is *trained* in how to carry out a task; he or she is *educated* in the principles of why something occurs or why they are performing an action. The operating procedures are used mostly in training, not education.

Training can be integrated with the modular concept of operating procedures. Along with the Action and Response columns, there can be two training columns: one of them called "Why?" and the other "Knowledge, Skills and Aptitudes (KSA.)" The "Why?" column provides an explanation and so is education rather than training. KSA information can be placed in the title block of each module. However, it can be excessively formal and conflicts with the philosophy of minimalist writing.

One way of linking procedures with training is to provide a training video to match each step of the procedures. The video would show an experienced operator actually working through the steps in the appropriate module. He would describe what he was doing for each of the steps, and the whole process would be recorded on a video tape. The tape would then be kept in the control room so that the operators could review it at any time.

This provides the operators with discursive and background material that is intentionally left out of their manual (which is written in a minimalist style.) The experienced operator could use the video to describe incidents that have occurred in the past relating to this step and he could also use it to show visually any difficulties or special features of the operation. This video would also provide a fine foundation for a trouble-shooting program.

BASIC/GENERIC TRAINING

Basic training is what an operator has to have before he is permitted to carry out any task, no matter how simple, on his own. The nature of the basic training will naturally vary from plant to plant. However, a program such as the one described below is quite representative.

- *Safety Policies*. This will principally be "hard hat" safety, rather than an understanding of process safety management issues.

- *Emergency Response*. The new operator will be made familiar with the plant emergency systems, and his role in the response, should an emergency occur.

- *Company Structure*. This part of the training will provide an overview of the different departments at the site, and what their responsibilities are. It will also explain the lines of authority, both in normal operation and at times of emergency.

- *Physical Movement*. This includes moving around the plant. Issues such as climbing ladders, buddy systems and the lifting of heavy equipment will be covered.

- *Special Safety Equipment*. The operator will be introduced to special safety equipment, such as self-contained breathing apparatus (SCBA) and chemically-resistant clothing, and be trained in its use.

- *Coordination with Other Departments*. The role of other departments, particularly maintenance, instrumentation and safety, will be explained. The way in which operations interacts with these departments, and who is responsible for what, is described. A particularly important part of this training is a description of the process for issuing work permits.

PLANT TRAINING

Once an operator has received his basic orientation, the next step is to provide him with plant training. The goal is to familiarize him with the processes, how they work, what they make, and how they inter-act with one another. This part of the training program will be even more plant and company-specific than the orientation. The elements outlined below are representative.

- *Plant or Refinery Overview.* This provides the trainee with an understanding of the overall process.

- *Major Equipment.* This section of the training will provide an overview of how to equipment items work, and how to operate them. The equipment covered will typically include pumps, heat exchangers, distillation columns, compressors and valves. Generic procedures will include items such as starting pumps, opening and closing valves and catching samples.

- *Instrumentation.* The trainee will be provided with an explanation of the plant's instrumentation, computer control systems and emergency shut down systems. The principles of measurement and control, including proportional, differential and integral controllers, will be explained.

- *Regulations.* The trainee will be introduced to the more important safety and environmental regulations, and told what he is expected to do to conform to those regulations.

INITIAL START-UP TRAINING

When the plant is new, all of the operators and supervisors require training. This training program can be included in Volume V — The Initial Start-Up Manual.

On the Job Training (OJT)

Once the operator has gone through the orientation and plant training steps, he can be put on a unit as a trainee operator. Initially, he will not be permitted to carry out even the simplest assignments on his own: everything that he does will have to be properly supervised by a senior operator. It is useful if the trainee can work a shift rotation that is different from the normal shift crews; that way he will be exposed to as many inputs and techniques as possible.

Computer-Based Training

Computer-based training is widely used in other industries, particularly aerospace and nuclear power to increase the effectiveness and efficiency of the training program. It is expensive to install, but is credited as being a significant factor in a 10% increase in the availability of French nuclear power plants over a ten year period [29]

Dynamic simulators were originally developed to model time-dependent processes, such as the change in pressure inside a reactor during a runaway reaction. As computers have become more powerful, these models can be used for inter-active training. This can be very useful because it allows the operators to handle all types of emergency or equipment problems.

Testing

At every step in the training process, the operator must be tested against some objective standard, with a clearly defined measure as to the distinction between passing and failing.

The tests can be very formal, such as multiple-choice answers for tests in a book. They can also be informal. For example, many supervisors include in their tests a simple walk around the unit with the trainee operator. During the walk around, the operator is asked a series of questions and he is meant to answer by showing what he would actually do in a particular situation.

One company had a number of very simple tests, most of which were known to the trainees ahead of time. For example, the trainees would be asked to draw from memory on a blank piece of paper a plot plan of their area, with the location of all safety and fire-fighting equipment pointed out.

Each company must decide on its policy concerning "grandfathering," i.e. whether it is acceptable for a very experienced operator who has not participated in the most recent training program, and who has not taken the tests associated with that program, to avoid having to take the test on the grounds that his experience makes it unnecessary. This can be a very delicate topic; if an operator has worked on a unit for a long time, the idea of being tested can be very threatening. Yet, it might be that management has to insist on his taking the test, because it could be, that, in spite of his experience, the operator really is not properly trained, especially in emergency operations.

Certification

Once an operator has successfully completed his training, and has worked in an area for a specified amount of time, he may be certified for the tasks for which he was trained. The exact meaning of the word certification will vary from company to company, and — particularly in unionized plants — the word may have legal implications. Broadly, however, certification simply means that an operator has participated in a properly designed and delivered training program, and that he has passed the tests that go with the program. Although there is some industry standardization, operator certification is something that each company sets up on its own; it is not a nationally accepted qualification such as a professional engineers license.

Contractors

Many accidents in process plants have involved contract workers performing maintenance duties. The purpose of this element is to require contract workers to be trained and managed to the same standard as the company's own employees.

Generally, contractors are not involved in the actual operation of a plant. Therefore, this element of the regulation is not likely to be important when writing operating procedures. However, the Permit to Work procedures (located in Volume II) will often involve a relationship with contract workers. This can be a problem because the contracting company may have its own set of procedures, and these need to be coordinated with the operating company's procedures.

Also, the standard requires that contractors be trained in the work that they are performing. This implies the existence of procedures, otherwise a training program cannot be developed.

PRE-STARTUP SAFETY REVIEW (PSSR)

Whenever a new project, such as a plant expansion, is carried out, it is important that the people who are responsible for running the plant assure themselves that they have been provided with a plant that is safe to operate. The Prestartup Safety Review element of the regulation provides them with a mechanism to carry this out.

This is particularly useful if the project is running over budget and/or behind schedule. When this occurs, the project managers will be tempted to cut corners. This element of the regulation, if properly carried out, gives the people who are to operate the unit the power to make sure that safety standards have not been allowed to slip. Part of the PSSR review will be to ensure that the operating procedures are correct and complete.

MECHANICAL INTEGRITY

This section describes the maintenance procedures for which the operator is responsible. It also describes the operator/maintenance interface, particularly the way in which hot work permits are handled (although this may also described in Volume II).

The goal of this element is to ensure that mechanical equipment does not fail such as to cause an accident. This element requires that all equipment in the plant be constructed according to the pertinent code, and that it be inspected on a regular basis. Once more, OSHA does not specify which codes are to be followed, or what the inspection methods and frequencies should be; that is up to the plant management. For many older plants, this element is often expensive. Its execution requires considerable manpower and the use of sophisticated equipment.

Although the Mechanical Integrity element is usually thought of in terms of checking equipment, the standard actually stresses the importance of procedures and training.

The nature and size of this section will depend largely on the extent to which the operators are responsible for maintenance. Some of the more important differences between operating and maintenance procedures are as follows.

- *Graphics*. Maintenance procedures make extensive use of equipment drawings, including cut-aways, exploded views and photographs. They rarely use process-oriented graphics, such as P & IDs or Iconic Flow Diagrams.

- *Information*. Maintenance procedures are likely to need access to large amounts of information on items such as model numbers, types of lube oil and special tools that may be needed.

- *Sequencing*. Maintenance procedures are not likely to need much branching. They are typically a straight sequence of instructions.

Organization of a mechanical integrity program is likely to follow a sequence such as is shown below.

1. Develop a program that meets the 1910.119 requirements (if necessary). This includes the control of material, equipment and supplies.
2. Identify and list all covered equipment.
3. Develop written procedures for mechanical integrity.

4. Train employees responsible for mechanical integrity.
5. Implement an Inspection and Testing program.
6. Correct any identified deficiencies.

HOT WORK

The goal of this element is to make sure that companies must have proper hot work permits (and other permit to work procedures such as first-break of lines, vessel entry and lockout/tagout). Most companies are already do a good job with this element; however it is important not to allow complacency to set in. Setting up a PSM program is a good opportunity to review policies, controls and to check out what is actually occurring in the field.

MANAGEMENT OF CHANGE

CONTROL OF CHANGE

PSM is about the control of change. Therefore the Management of Change component of PSM is fundamental to the program's success or failure because most accidents will arise from the consequences of an unintended change. Many accidents, such as the one that occurred at Flixborough in the United Kingdom, are attributable to badly-managed changes that caused the system to fail in some way. Additional examples of change leading to unforeseen consequences are provided by Sanders [30].

Management of Change is not intended to prevent change from taking place. Change is a fact of life. Indeed, on a large plant, conditions change almost daily. Equipment is changed, organizations are modified, new people are brought on board, operating parameters are extended; the list of possible changes is almost endless. Therefore, it is essential that management put in place a mechanism for (a) identifying change, and (b) ensuring that all the proper controls are in place to ensure that the change does not lead to an accident.

If a company is instituting a PSM program, Management of Change is the first element that should be put in place (along with

P & IDs). This is because, if it is not in place, as the other elements are instituted, they will generate changes that will not be properly documented or analyzed.

Management of Change and Operating Procedures are closely connected because so many plant changes will trigger a change in the procedures. The reason for this is that the procedures describe the person-machine interface. Therefore, change to either equipment or to the organization will affect this interface, and so the procedures will have to be updated.

Management of Change can be viewed as a four-step process, as described below.

1. *Initiation of Change.* Change can be initiated by any person on the plant. Indeed, all employees should be encouraged to suggest changes and improvements. In the case of operating procedures, if an operator wants to suggest a change, all that he need do is photocopy the pertinent pages, hand-write the proposed changes on to those pages, and enter them into the management of change system.

2. *Review.* Major changes will be subjected to a process hazards review before being implemented. In practice, this will usually be a HAZOP, but it does not have to be so.

3. *Implementation.* Once a change has been reviewed and accepted, it can be implemented.

4. *Communication and Training.* Once the change has been initiated, and the procedures updated, it is essential that all the operators, and any other affected personnel, be informed as to the nature of the change and that they be trained in how to run the unit under these new conditions.

Management of Change and Operating Procedures

It has already been stressed that a major problem for many companies is keeping their manuals up to date. This is often harder than preparing a good first edition. The procedures represent the interface between the equipment and the people responsible for running it. Consequently, a large percentage of both equipment and organizational changes are likely to impact the procedures. The procedures should also be changed as a result of the operators making suggestions as to how they can be improved.

One way of reducing the problems with this is to make sure, as has already been stressed in previous chapters, that the manual be organized and laid out so that the act of updating is made as easy as possible. This is the reason for stressing the concept of modular design: modules can be updated, added, removed and re-located with minimal effort.

Updates also require that a functioning management of change system be in place. For example, if an operator wants to make a suggestion as to how a procedure could be improved, he can simply photo-copy the pertinent page(s) and hand write on it the changes that he would like to see made. These suggestions can then be fed into the management of change system. Operating procedures are only one component of this.

It is very difficult to predict how often the MOC process has to be executed. Many larger facilities are finding that, once the system is settled down, a significant change is required to the operating manual is required about once per week. Clearly, however, this will depend very heavily on the type of operations that are taking place.

Research and pilot facilities have a lot of trouble with this topic. By intent, they are constantly changing conditions, and they are frequently changing conditions into areas where there is no operating experience. Therefore, they are not within the bounds of experience and so MOC applies. This is very difficult to execute because of the degree to which it slows down progress.

Use of Blank Process Flow Diagrams

Once a plant has been running for more than a few months, the operating conditions may well be quite different from what are shown in the design PFDs. There are two reasons for this. The first is that the original estimates may have been off, and actual operating conditions are different from what were anticipated. The second reason is that plant conditions will be deliberately changed in order to improve safety, production or yields.

Given that the basic process flow information is so important, it is regrettable that the original PFD becomes relatively uninformative after a short time. One way around this is for the unit supervisor to fill in a blank PFD with current conditions. This can be modified for different sets of operating conditions.

Replacement In Kind

The standard excludes the need for management of change procedures when there is a "replacement in kind." For mechanical items, this is usually quite straightforward. If a gasket is to be replaced, then, if the replacement is identical, the there is no need for Management of Change review. If the new gasket is different, however, such a review is required.

In the case of operating procedures, there will never be a true replacement in kind — any modification will be a change. There are, however, some changes that could be viewed as being equivalent to replacement in kind, and that would not trigger the Management of Change standard. Examples of these changes are:

1. Correction of misspellings, typographical errors, illegible words and grammatical corrections.

2. Clearly erroneous data. For example, on a plant built in Europe, any reference to degrees Fahrenheit can probably be assumed to be degrees Celsius (however, caution is required — if the licensor is American, this may not be "just a typo.") This is one reason why it is so important to always use units.

3. Blanks and missing sections. If equipment numbers or locations are missing they can be filled in right away.

Cross-references that have changed. If a module changes its title, for example, all the other modules that call it will have to be modified.

Incident Investigation

The purpose of this element is to provide management with a wake-up call. Minor incidents and near-misses, no matter how trivial, show that management is not in full control of their facility.

Minor incidents are also harbingers of more serious accidents. There are two reasons for this.

First, a serious accident usually requires the simultaneous occurrence of two or more independent events (the AND Gate in Fault Tree parlance [31].) Such a simultaneous occurrence is unlikely, but the occurrence of one of the independent events by itself is much more likely. Therefore, it is much more likely to show up in the incident reporting system. So, for example, the explosion of a pressure vessel might require three independent events to occur: (1) the pressure increases to an unsafe level, (2) the operator fails to take corrective action, and (3) the relief valve fails to open. If, for sake of argument, the respective probabilities of these events are 0.01 yr^{-1} , 0.1 and 0.05, i.e. once in 100, once in ten years and once in twenty years respectively, the probability of the vessel exploding is 0.00005 yr^{-1}, i.e. once in a 20,000 years. (Note that events 2 and 3 are dimensionless, because they are probabilities of *occurrence*, not *frequency rates.*) Therefore, the overall probability of this serious accident is very low, but the probability of any one of these events occurring is quite significant, and will probably show up within five years of the incident investigation reporting system being implemented.

Thus, the incident investigation system provides a warning that the integrity of a redundant safety system has been violated, or, to use a military analogy, one line of defense has been penetrated, thus increasing the probability of an enemy breakthrough. This is illustrated in Figure 29.

Figure 29

Fault Tree Example

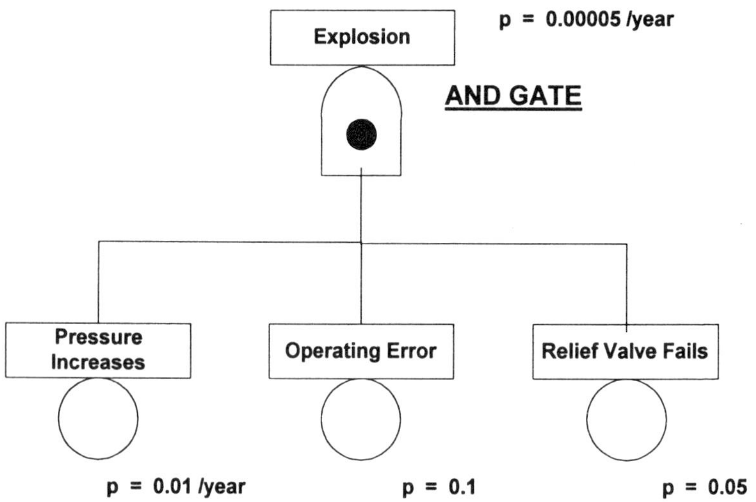

The second reason for stressing incident investigation is simply that it provides real-life evidence as to where the hazards of the process are to be found. All other methods, such as HAZOPs and fault tree analysis, are theoretical and/or analytical. Incident investigation provides a reality check.

If the Incident Investigation element is to work properly, it is, of course, absolutely essential that all personnel feel free to report incidents and near-misses without fear of retribution either to themselves or, and this may be more important, to their co-workers. In fact, the mood should be completely turned around; those people who report accidents and near-misses should be praised for doing so, not reprimanded for doing something that could have caused an accident.

In order to ensure the integrity of the incident reporting system, the employee who first reports the incident or near miss should be kept fully informed as to the status of the report and management's follow-up. This is yet another form of employee participation.

EMERGENCY PLANNING AND RESPONSE

The regulation requires that the plant have an emergency response plan in place. In practice, many of the requirements of this section are covered by other OSHA regulations. Naturally, many of the requirements of this element are met by Volume III of the operating procedures (the Emergency Response Plan.) Other types of OSHA emergency response standards are shown in Table 17 (and in Table 9 in Chapter 4.)

Table 17

OSHA Emergency Response Regulations

Regulation	Text
29 CFR 1910.119 **Process Safety Management** *Emergency Planning and Response*	The employer shall establish and implement an emergency action plan in accordance with the provisions of 29 CFR 1910.38(a). Note: 29 CFR 1910.120 (a), (p) and (q) may also be applicable
29 CFR 1910.38 **Employee emergency plans and fire prevention plans**	(a) *Emergency action plan* (2) *Elements*. The following elements, at a minimum, shall be included in the plan: (i) Emergency escape procedures and emergency escape route assignments; (ii) Procedures to be followed by employees who remain to operate critical plant operations before they evacuate; (iii) Procedures to account for all employees after emergency evacuation has been completed. (iv) Rescue and medical duties for those employees who are to perform them; (v) The preferred means of reporting fires and other emergencies; and (vi) Names or regular job titles of persons or departments who can be contacted for further information or explanation of duties under the plan.

Table 17 (continued)

29 CFR 1910.120	(p)(8) Emergency response program (i) An emergency response plan shall be developed and implemented by all employers. Such plans need not duplicate any of the subjects fully addressed in the employer's contingency planning required by permits, such as those issued by the U.S. Environmental Protection Agency, provided that the contingency plan is made part of the emergency response plan. The emergency response plan shall be a written portion of the employer's safety and health program required in paragraph (p)(1) of this section. Employers who will evacuate their employees from the work-site location when an emergency occurs and who do not permit any of their employees to assist in handling the emergency are exempt from the requirements of paragraph (p)(8) if they provide an emergency action plan complying with §1910.38(a) of this part.
	(ii) Elements of an emergency response plan. The employer shall develop an emergency response plan for emergencies which shall address, as a minimum the following areas to the extent that they are not addressed in any specific program required in this paragraph. (A) Pre-emergency planning and coordination with outside parties. (B) Personnel roles, lines of authority, and communication. (C) Emergency recognition and prevention. <Some additional items follow>

Table 17 (continued)

	(iv) Procedures for handling emergency incidents. (A)(1) Site topography, layout and prevailing weather conditions. (A)(2) Procedures for reporting incidents to local, state and federal government agencies. (B) The emergency response plan shall be compatible and integrated with the disaster, fire and/or emergency response plans of local, state, and federal agencies. (iv)(C) The emergency response plan shall be rehearsed regularly as part of the overall training program for site operations. (iv)(D) The site emergency response plan shall be reviewed periodically and, as necessary, be amended to keep it current with new or changing site conditions or information.
	(q)(1) An emergency response plan shall be developed and implemented to handle anticipate emergencies prior to the commencement of emergency response operations. The plan shall be in writing and available for inspection and copying by employees, their representatives and OSHA personnel. <Some exceptions follow>

Table 17 (continued)

	(q)(2) The employer shall develop an emergency response plan for emergencies which shall address, as a minimum, the following to the extent that they are not addressed elsewhere: (i) Pre-emergency planning and coordination with outside parties. (ii) Personnel roles, lines of authority, training and communication. (iii) Emergency recognition and prevention. (iv) Safe distances and places of refuge. (v) Site security and control. (vi) Evacuation routes and procedures. (vii) Decontamination. (viii) Emergency medical treatment and first aid. (ix) Emergency alerting and response procedures. (x) Critique of response and follow-up. (xi) PPE and emergency equipment.

COMPLIANCE AUDITS

There are two types of audit. The first is the formal type of audit that would be performed by OSHA during an inspection, or after an accident. Similar to this is an audit that is internally funded, but that is handled very formally. An issue with this type of audit is that, once the results have been documented, the company *must* respond the findings.

The second type of audit is less formal, but more frequent. If informal audits are carried out routinely, the results of the formal audit should never be a surprise. The use of the 'Response' column in the instructions in Volume I is an informal audit process.

The term *Verification and Validation* (V & V) is sometimes used by auditors, particularly in the nuclear power industry [32]. In the context of operating procedures, verification refers to the process of making sure that they are technically accurate and follow the writing rules established in the design basis for the procedures. Validation concerns usability, operational correctness and workability.

Tables 18 provides an example of a short audit protocol for operating procedures.

Table 18

Sample Audit Protocol - Operating Procedures

	Concern or Question	Finding	Verif.
1.0	**OVERVIEW**		
1.1	What are the Goals of the Procedures?		
1.2	Are the Operating Procedures Written Down?		
1.3	Are employees involved in writing and maintaining operating procedures?		
1.4	Are employees involved in the preparation of procedures in any other manner?		
1.5	Are all sections of the plant provided with written procedures?		
1.6	Are the operating procedures consistent with the process safety information?		
2.0	**PROCESS SAFETY INFORMATION**		
2.1	Are operating limits provided?		
2.2	Are maximum inventory limits provided?		
2.3	Are consequences of deviations from operating limits explained?		

Compliance Guides

To supplement the regulation itself, OSHA inspectors are provided with a compliance guides. These provide assistance on how to conduct an inspection of a facility for compliance with 1910.119. They suggest looking for items such as the following:

1. A statement as to whether or not the plant meets that particular requirement
2. A list of the documents reviewed and personnel interviewed (preferably by job title rather than personal name)
3. A description of any identified deficiencies
4. A description of work in hand that addresses the deficiencies
5. Any recommendations

Table 19 shows the Compliance Guide that OSHA inspectors will follow [33]. This is dated September 13, 1994. The formatting is slightly different from the original.

Table 19

OSHA Compliance Guide for Operating Procedures

<table>
<tr><td colspan="3">I. PROGRAM SUMMARY
The intent of this paragraph is to provide clear instruction for conducting activities involved in covered processes that are consistent with the process safety information. The operating procedures must address steps for each operating phase, operating limits, safety and health considerations, and safety systems and their functions.</td></tr>
<tr><td colspan="3">II. QUALITY CRITERIA REFERENCES</td></tr>
<tr><td colspan="3">A. 1910.119(f) D. 1910.1000
B. 1910.120 E. 1910.1200
C. 1910.147</td></tr>
<tr><td>III. VERIFICATION OF PROGRAM ELEMENTS</td><td>Criteria Reference</td><td>Met Y/N</td></tr>
<tr><td>A. Records Review

1. Do written operating procedures exist for each covered process? Do the procedures provide clear instructions for conducting activities safely?

FIELD NOTE REFERENCES:</td><td>.119(f)(1)</td><td></td></tr>
</table>

Table 19 (continued)

2. Do the operating procedures address, as a minimum, steps for each operating phase, including: • Initial start-up? • Normal operations? • Temporary operations? • Emergency shutdowns? I. Conditions requiring emergency shutdown? II. Assignment of shutdown responsibility to qualified operators? III. Emergency operations? IV. Normal shutdown? V. Start-ups following a turnaround or emergency shutdown? FIELD NOTE REFERENCES:	.119(f)(1) (i)	
3. Do the operating procedures include operating limits that outline consequences of process deviation and steps required to correct or avoid deviations? FIELD NOTE REFERENCES:	.119(f)(1) (ii)	

Table 19 (continued)

4. Have safety and health considerations been included in the operating procedures? Do they include at a minimum: • Properties of, and hazards presented by, chemicals used in the process? • Precautions necessary to prevent exposure, including engineering controls, administrative controls, and personal protective equipment? • Control measures to be taken if physical contact or airborne exposure occurs? • Quality control for raw materials and control of hazardous chemical inventory levels? • Any special or unique hazards? FIELD NOTE REFERENCES:	.119(f)(1) (iii)	
5. Are safety systems and their functions included in the operating procedures? FIELD NOTE REFERENCES:	119(f)(1) (iv)	
6. Are the operating instructions consistent with the process safety information? FIELD NOTE REFERENCES:	.119(f)(1)	
7. Are the operating procedures readily accessible to employees who work in or maintain a process? FIELD NOTE REFERENCES:	.119(f)(2)	

Table 19 (continued)

8. Are the operating procedures reviewed as often as necessary to assure that they reflect current operating practice? Are they certified annually by the employer that they are current and accurate? Do they reflect current operating practices that have resulted from changes in: • Process chemicals? • Technology? • Equipment? • Facilities? FIELD NOTE REFERENCES:	.119(f)(3)	
9. Have safe work practices been developed and implemented for employees and contractors to control hazards during operations such as: • LockOut/tagout? • Confined space entry? • Opening process equipment or piping? • Control over entrance into a facility by maintenance, contractor, laboratory or other support personnel? FIELD NOTE REFERENCES:	.119(f)(4)	
B. On-site Conditions 1. Does observation of a representative sample of processes indicate that the written operating procedures are being implemented? FIELD NOTE REFERENCES:	119.(f)(1)	

Table 19 (continued)

2. Does observation of a representative sample of processes indicate that the written operating procedures are readily accessible to employees who work or maintain a process? FIELD NOTE REFERENCES:	119.(f)(2)	
3. Does observation of a representative sample of processes indicate that operating procedures reflect current practice, including changes that result from process chemicals, technology, equipment, and facilities? (Observe to see if actual procedures match the written operating procedures.) FIELD NOTE REFERENCES:	.119(f)(3)	
4. Does observation of representative operations indicate that safe work practices have been implemented for company and contractor employees? Do work practices include, where appropriate: • Lockout/tagout? • Confined space entry? • Opening process equipment or piping? • Control over entrance into a facility by maintenance, contractor, laboratory, and other support personnel? FIELD NOTE REFERENCES:	.119)f)(4)	

Table 19 (continued)

C. Interviews 1. Based on interviews with a representative number of operators, are the written operating procedures implemented for each covered process? FIELD NOTE REFERENCES:	.119(f)(1)	
2. Based on interviews with a representative number of operators, do operating procedures provide clear instructions for safely conducting activities? (Specifically, ask for conditions requiring emergency shutdown, the operating limits of a particular process or item or equipment, what might occur if a deviation from those limits should take place, steps to avoid the deviation, and precautions necessary to prevent exposure to hazardous chemicals.) FIELD NOTE REFERENCES:	.119(f)(1)	
3. Based on interviews with a representative number of employees who work in or maintain a process, are the operating procedures readily accessible? FIELD NOTE REFERENCES:	.119(f)(2)	
4. Based on interviews with a representative number of operators and maintenance employees, do the operating procedures reflect current operating practice? FIELD NOTE REFERENCES:	.119(f)(3)	

Additional audit protocols for operating procedures are provided by the Nuclear Regulatory Commission [34].

EXAMPLES OF DETAILED AUDIT FORMS

Table 20 provides some representative detailed audit forms. In addition to posing audit questions, these forms provide some guidance as to what to look for. The "Guidance and Discussion" section provides the auditor with practical hints on what to look for that would provide him or her with clues as to how well that question is being addressed.

Table 20

Detailed Audit Forms

Audit Question 1.1	WHAT ARE THE GOALS OF THE PROCEDURES?
Auditor's Name and Company	
Date of Audit and Report	
Guidance and Discussion	Regulations usually focus on safety. Find out from the client to what extent he/she expects the procedures to address operational and quality issues. Adjust this audit appropriately. For example, the plant may be preparing for ISO 9000 certification. This will affect the design of the procedures. Some companies refer to their operating procedures as the "Training Manual." Determine the relationship between procedures and training.
Persons Interviewed	
Documents Reviewed	
Findings	
Work In Progress	
Recommendations or Comments	

Table 20 (continued)

Audit Question 1.2	ARE THE OPERATING PROCEDURES WRITTEN DOWN?
Auditor's Name and Company	
Date of Audit and Report	
Guidance and Discussion	In some plants, the procedures may be "written" directly on the DCS (Distributed Control System). If this is the case, check that the panels can be accessed quickly and that they are easy to read. It is not clear if electronic writing of this sort will meet OSHA requirements.
Persons Interviewed	
Documents Reviewed	
Findings	
Work In Progress	
Recommendations or Comments	

Table 20 (continued)

Audit Question 1.3	ARE EMPLOYEES INVOLVED IN WRITING AND MAINTAINING THE OPERATING PROCEDURES?
Auditor's Name and Company	
Date of Audit and Report	
Guidance and Discussion	Check this with randomly selected employees. Do this on the off-shifts also. Include both experienced and inexperienced operators.
Persons Interviewed	
Documents Reviewed	
Findings	
Work In Progress	
Recommendations or Comments	

Table 20 (continued)

Audit Question 2.1	**ARE OPERATING LIMITS PROVIDED?**
Auditor's Name and Company	
Date of Audit and Report	
Guidance and Discussion	This is often a major weakness in operating procedures. The operators should be provided with at least six values for each important variable: i) The normal value ii) The normal upper range above which the plant is in "trouble." iii) The normal lower range. iv) The upper emergency level. v) The lower emergency level. vi) Alarm/interlock values. Emergency values may not be needed in all cases. For example, "low reactor temperature" is often not an emergency problem.
Persons Interviewed	
Documents Reviewed	
Findings	
Work In Progress	
Recommendations or Comments	

Table 20 (continued)

Audit Question 2.2	ARE MAXIMUM INVENTORY LIMITS PROVIDED?
Auditor's Name and Company	
Date of Audit and Report	
Guidance and Discussion	Check that maximum inventory limits are written down, particularly for hazardous chemicals. This is one of the most important parts of Audit Question 2.1. Accidents commonly occur because vessels are over-filled or because liquid seals are allowed to blow by.
Persons Interviewed	
Documents Reviewed	
Findings	
Work In Progress	
Recommendations or Comments	

Table 20 (continued)

Audit Question 2.3	ARE CONSEQUENCES OF DEVIATIONS EXPLAINED?
Auditor's Name and Company	
Date of Audit and Report	
Guidance and Discussion	The procedures should explain the consequences of both "trouble" and "emergency." This includes operational, environmental and safety issues.
Persons Interviewed	
Documents Reviewed	
Findings	
Work In Progress	
Recommendations or Comments	

Table 20 (continued)

Audit Question 2.4	**ARE THE STEPS TO AVOID DEVIATION EXPLAINED?**
Auditor's Name and Company	
Date of Audit and Report	
Guidance and Discussion	These should be attached to the "consequences" statement in the procedure. It includes the use of safety systems such as fire monitors and hydrocarbon sniffers.
Persons Interviewed	
Documents Reviewed	
Findings	
Work In Progress	
Recommendations or Comments	

Table 20 (continued)

Audit Question 2.5	IS THERE A CLEAR STATEMENT AS TO WHAT CONSTITUTES AN EMERGENCY?
Auditor's Name and Company	
Date of Audit and Report	
Guidance and Discussion	It is important to distinguish between "trouble," which usually leads to economic losses, and emergencies which are fast-moving and which can lead to loss of life.
Persons Interviewed	
Documents Reviewed	
Findings	
Work In Progress	
Recommendations or Comments	

Table 20 (continued)

Audit Question 3.1	**IS INITIAL START-UP EXPLAINED?**
Auditor's Name and Company	
Date of Audit and Report	
Guidance and Discussion	The OSHA regulation draws a distinction between initial startup and subsequent plant starts. In practice, the two will often be very similar. Initial startup will only be used when the plant is brand new or when it is coming off a major turnaround.
Persons Interviewed	
Documents Reviewed	
Findings	
Work In Progress	
Recommendations or Comments	

Table 20 (continued)

Audit Question 3.2	ARE NORMAL OPERATIONS EXPLAINED?
Auditor's Name and Company	
Date of Audit and Report	
Guidance and Discussion	There is an important distinction here between plants that operate a steady-state operation and those that are primarily batch. For steady-state plants, the normal procedures are often quite limited. They consist primarily of checking to make sure that there are no deviations from operating goals. In the case of batch plants, the normal operating procedures are essentially the same as the startup and shutdown procedures. Sampling is sometimes a high hazard normal operation. Verify that all sample operations are properly written up and that instructions as to what to do if there is an uncontrolled leak from a sample point are provided.
Persons Interviewed	
Documents Reviewed	
Findings	
Work In Progress	
Recommendations or Comments	

Table 20 (continued)

Audit Question 3.3	**ARE TEMPORARY OPERATIONS UNDERSTOOD AND ARE PROCEDURES WRITTEN FOR THEM?**
Auditor's Name and Company	
Date of Audit and Report	
Guidance and Discussion	This is often an area of weakness. In many cases, temporary operations have to be done quickly and they do not take much time (for example, a hose run-around a leaking valve). Therefore, the plant has a tendency to skip writing procedures. All temporary operating procedures should be kept on file. Ask to look for this file. In particular, check that the temporary procedures include their own termination date.
Persons Interviewed	
Documents Reviewed	
Findings	
Work In Progress	
Recommendations or Comments	

Table 20 (continued)

Audit Question 3.4	IS EMERGENCY SHUTDOWN UNDERSTOOD AND EXPLAINED?
Auditor's Name and Company	
Date of Audit and Report	
Guidance and Discussion	Generally, this will be covered in the *Emergency Operating Procedures* (next section). Ensure that the procedures make a distinction between the two, particularly in those cases where the best emergency response is to keep the plant running. Check that the circumstances that require abandoning the plant and shutting down are properly described.
Persons Interviewed	
Documents Reviewed	
Findings	
Work In Progress	
Recommendations or Comments	

Table 20 (continued)

Audit Question 3.5	ARE EMERGENCY OPERATIONS UNDERSTOOD AND EXPLAINED?
Auditor's Name and Company	
Date of Audit and Report	
Guidance and Discussion	The procedures should be short. The operators should have them memorized. Many plants draw a distinction between Operating Procedures and the Emergency Response Plan. Ensure that this distinction is properly understood and explained.
Persons Interviewed	
Documents Reviewed	
Findings	
Work In Progress	
Recommendations or Comments	

TRADE SECRETS

OSHA recognizes that many processes are proprietary, and that the PSM program may lead to sensitive information being released. This section of the regulation mandates that companies make required information available to their employees, regardless of the information's trade secret status. Many plants, particularly refineries, do not have their own proprietary technology. However, they often operate licensed processes from other companies. This element of the regulation provides protection to the licensors.

Although the trade secrets section of the regulation provides important legal coverage, it has little technical impact, and is not discussed further in this book.

OTHER REGULATIONS AND STANDARDS

This section provides a brief overview of the way in which regulations and standards are structured.

THE EPA RISK MANAGEMENT PROGRAM (RMP)

EPA's Risk Management Program is similar to OSHA's PSM standard. The major differences are: "those mandated by the CAA, such as the hazard assessment (offsite consequence analysis, the five-year accident history), the emergency response requirements, registration, and the RMP submission to the Board . . ." and "for most prevention program elements, facilities that are in compliance with OSHA's process safety management standard will not need to do anything different or create different onsite documentation" [35]. The RMP standard covers a larger number of processes, and has more reporting requirements. The element of this regulation concerning operating procedures is quoted in Table 21 (from the above reference).

Table 21

The EPA Standard for SOPs

The results of the process hazard analysis, information developed during the design of a process, and industry and facility experience combine to define the proper way to conduct operations and maintain equipment. SOPs describe the tasks to be performed by the operator, the operating parameters (e.g., temperature, pressure) that must be maintained, and safety precautions needed for both operations and maintenance activities. SOPs must specify the consequences of deviations from safe operating limits (e.g., if the safe operating temperatures are between 100 and 150C, the SOPs should indicate what happens if the temperature is above or below those limits). Written SOPs provide a guide to safe operations in a form that can be used by employees. Lack of SOPs and inadequate SOPs have been implicated in a number of catastrophic accidents. For example, improper maintenance procedures have been blamed for a release and explosion at a facility in New Castle, Delaware, in 1980, which killed six people, injured 27 others, and caused more than $63 million in property damage to the facility.

SOPs, which define proper steps to take in these emergency situations, provide a quick source of information that can prevent or mitigate the effects of accidents. SOPs also provide workers and management a standard against which to assess performance; the procedures clarify for both operators and supervisors how operations should be carried out at the facility. Application of SOPs can result in more cost-effective operations by ensuring that operators adhere to procedures that maximize both the safety and efficiency of the process.

Table 21 (continued)

EPA is proposing that each facility develop written SOPs for each process and operation involving the regulated substance above the threshold. The SOPs would include instructions on steps for each operating phase (e.g., initial startup, normal operation, emergency shutdowns, normal shutdowns, emergency operations), operating limits, safety and health considerations, and safety systems. The facility would also be required to provide for control of hazards during operations involving lockout/tagout, confined space entry, and operating process equipment or lines. The facility would also need SOPs to control entrance to the facility by support personnel. The level of detail included in the SOP should be appropriate for the operation covered. For example, instructions for proper storage of chemicals may be relatively brief, while procedures for routine startup of a complex process may require considerable detail to ensure that each action required is detailed and explained. EPA emphasizes that SOPs should be usable by the operators in running the process; that is, the SOPs should be written in a language and at a level appropriate for the operators.

The language of the above regulation is different in detail from the OSHA standard, but the intent is very similar. Differences of contents are very small. For example, the reference to "control entrance to the facility.") If procedures are written in the manner outlined in this book, they should meet the above requirements in most cases.

States

Although OSHA's PSM standard is used in this book to provide guidance as to how regulations are structured, and how they should be followed, there are other PSM regulations within the United States that are applicable, mostly from individual states. About 25 states have adopted their own version of PSM. As long as the requirements of a state version are at least as demanding as the federal version, it is the one that has to be followed. For example,

California's PSM regulation does not exempt atmospheric storage tanks from its requirements, even though the federal regulation does. Therefore, the state regulation takes precedence.

Some states also have risk management programs that do not correspond directly to 1910.119 but which nevertheless have strong similarities. These programs are addressed more towards protecting the public, whereas OSHA programs are concerned primarily with the workers. Examples of these state regulations include California's Risk Management and Prevention Program[36] and New Jersey's Toxic Catastrophe Prevention Act (TCPA). The New Jersey regulation for operating procedures is shown in Table 22.

Table 22

The New Jersey Standard for Operating Procedures

(a) The standard operating procedures shall be written in English in a manner understandable by the EHS [extraordinarily hazardous substance] operators and shall also be written in the language of fluency of EHS operators not fluent in English.

(b) A copy of the standard operating procedures shall be readily available to the EHS operators.

(c) The standard operating procedures shall include, but not be limited to:

1. Simplified process flowsheets and a process description defining the operation and showing flows, temperatures and pressures;
2. Procedures and conditions for normal operations;
3. A description of abnormal conditions, including the control and mitigating procedures to be followed to return to normal conditions;
4. A description of emergency conditions which could occur including the control and mitigating procedures to be followed to reduce the impact of emergency conditions;
5. Pre-start up procedures covering, at a minimum, testing for leak tightness prior to placing into EHS service;
6. Startup procedures including conditions to be maintained during startup;

Table 22 (continued)

7. Shutdown procedures including provisions for normal and emergency shutdown and details on the condition of equipment to be maintained after shutdown;

8. A description of the type, location and purpose of safety relief devices, interlocks and alarms with their respective activation points indicated;

9. Sampling procedures addressing apparatus and specific steps involved in the taking of samples;

10. Safety procedures related to each specific operation in the standard operating procedures;

11. Procedures to prepare EHS equipment for maintenance and inspection of maintenance work upon completion and prior to placement of equipment in EHS service;

12. Material safety data sheets or fact sheets for each EHS used in the operation;

13. Logsheets and checklists where appropriate to the operation;

14. A statement as to the number of EHS operators required to meet safety needs for each operation with requirements for shift coverage; and

15. A requirement that the EHS operator be in attendance at the EHS site, be able to acknowledge alarms and take corrective action to prevent an accident at all times during EHS handling, use manufacturing, storage, or generation, . . .

INDUSTRY STANDARDS

The regulations discussed above have to be followed; they are the law of the land. In addition, many industry groups have developed standards that represent good practice — they recommend that their members follow these standards on a voluntary basis. There are many groups of this type; some of the more important are mentioned below.

American Petroleum Institute Recommended Practice 750

API RP 750 [37] shares the same roots as 1910.119: both are based in a report by the Organization of Resource Councilors. Therefore, there are very strong similarities between the two. It is used primarily by refineries rather than chemical plants.

ISO 9000

The term ISO 9000 covers a group of quality standards that are widely used in Europe and that are gaining increasing acceptance in the United States. At the heart of an ISO 9000 program is the concept of procedures: all activities are carefully structured and defined and then formalized in an appropriate procedure.

The ISO standards do not, however, provide instructions as to how procedures are to be written. They are concerned with document control and with ensuring that the only procedures that are available are current and that they are present at the work-station or in the control room. Furthermore, ISO 9000 is concerned only with product quality; procedures in other areas, such as safety or energy-savings, are not within its scope.

REFERENCES

[1] Chiu, C. "Human-factors Engineering Key to Trip Reduction." *Power.* July 1988.

[2] Reason, J.T. *Human Error.* Cambridge: Cambridge University Press, 1990

[3] Geyer, T.A.W., L.J. Bellamy, J.A. Astley, and N.W. Hurst. "Prevent pipe failures due to human errors." *Chemical Engineering Progress.* November 1990.

[4] "Statistical Summary of Commercial Jet Aircraft Accidents." Boeing Commercial Airplanes, Seattle, WA. 1989.

[5] Pitblado, R.M., L.J. Bellamy, and T. Geyer. *Safety Assessment of Computer Controlled Process Plants.* Technica

[6] James, Geoffrey, "Artificial Intelligence and Document Processing". *Proceedings of the Fifth International Conference on System Documentation.* University of Toronto. June 1986.

[7] Sandler, Henry J., and Edward T. Luckiewicz. *Practical Process Engineering.* New York: McGraw-Hill, 1987.

[8] Weiss, Edmond H., *How to Write a Usable User Manual.* Philadelphia: ISI Press. 1985.

[9] Rasmussen. "Chemical Process Hazard Identification". *Reliability Engineering and System Safety.* 1989.

[10] Fawcett, H.H., and W.S. Wood. *Safety and Accident Prevention in Chemical Operations.* New York: John Wiley. 1987

[11] Gans, M., D. Kohan, and B. Palmer, "Systematize troubleshooting techniques". *Chemical Engineering Progress.* April 1991.

---

[12] Kletz, T.A. “Computers and HAZOP”. *The Chemical Engineer*. October 1988.

[13] Andow, P., and G. Ferguson. “Expert systems and chemical process safety”. Paper presented at the World Bank/AIChE/EPA Conference, in Washington D.C. February 1987.

[14] Checklist for Plant Completion. API 700. March 1972.

[15] Gunning, Robert. “New Guide to More Effective Writing in Business and Industry”. 1963.

[16] Strunk, William Jr., and E.B. White. *The Elements of Style.* New York: Macmillan Publishing. 1979.

[17] Lannon, John M. *Technical Writing.* Harper Collins. 1991.

[18] *The Chicago Manual of Style.* Chicago: University of Chicago Press. 1993.

[19] Brooks, Frank P. *The Mythical Man-Month.* New York: Addison-Wesley. 1975.

[20] Department of Energy Standard (for Trial Use). “Writer’s Guide for Technical Procedures.” U.S. Department of Energy, Office of Nuclear Energy. DOE/NE/SP-0001T. September 1991.

[21] Williges, Beverly H., and R.C. Williges, “Dialogue Design Considerations for Interactive Computer Systems”. *Human Factors Review*. 1984. 167-208.

[22] Tinker, Miles A. *Bases for Effective Reading.* Minneapolis: University of Minnesota Press. 1965.

[23] Danchak, M.M. “CRT Displays for Power Plants”. *Instrumentation Technology*, 23, 10. 1976: 29-36.

24 Horton, William K.. *Designing and Writing On-Line Documentation.* New York: John Wiley & Son. 1989.

25 Moore, David A. *Integrating Process Safety Management with ISO 9000.* 1992.

26 Moubray J., *Reliability Centered Maintenance.* Butterworth-Heinemann, Ltd. 1991.

27 Sutton, Ian S. "Integrated Management Systems Improve Plant Reliability." *Hydrocarbon Processing.* January 1995.

28 Brandes, R.L. "PSM auditing: Opportunity to improve reliability at the plant". *Chemical Process Safety Report.* November 1994.

29 Widen, William C., M.A. Managan, and B.A. Weiss. "Computer-Based Training: An Evolving Tool for Nuclear Training." *Nuclear Plant Journal.* 1988.

30 Sanders, R.E. "Small, Quick Changes Can Create Bad Memories". *Chemical Engineering Progress.* May 1992.

31 Sutton, Ian S. *Process Reliability and Risk Management.* New York: van Nostrand Reinhold. 1992.

32 Fleger, S.A. "The Role of Human Factors Engineering in Process Safety Management". Submitted to Occupational Health and Safety. October 24, 1991.

33 U.S. Department of Labor, Assistant Secretary for Occupational Safety and Health, Directorate of Compliance Programs. "29 CFR 1910.119, Process Safety Management of Highly Hazardous Chemicals — Compliance Guidelines and Enforcement Procedures." OSHA Instruction CPL 2-2.45A CH-1. September 13, 1994.

34 "NRC Inspection Manual — Operating Procedures." March 1994.

[35] "Risk Management Programs for Chemical Accidental Release Prevention". Proposed Rule. Environmental Protection Agency, 40 CFR Part 68. 1993.

[36] "Risk Management and Prevention Program Guidance". Governor's Office of Emergency Services, Advance Copy. State of California. June 1989.

[37] "Management of Process Hazards". *API Recommended Practice 750*, First Edition. American Petroleum Institute. 1990.

INDEX

—A—

—B—

—C—

—D—

—E—

—F—

—G—

—H—

—I—

—J—

—K—

—L—

—M—

—N—

—O—

—P—

—Q—

—R—

—S—

—Y—